高等职业教育土木建筑类专业新形态教材

建筑英语
（第2版）

主 编　郭二莹　姜　华
副主编　孙熙阳　刘　玥　田妲妮

北京理工大学出版社
BEIJING INSTITUTE OF TECHNOLOGY PRESS

内容提要

本书共10个单元，主要内容包括世界建筑史、中国古代建筑、建筑设计、建筑材料、建筑结构、建筑施工、房屋设备、室内装饰、园林设计以及生态与建筑。每个单元由Text A、Text B、Oral Practices和Translation Skills四部分组成。全书内容丰富、选材广泛、图文并茂、易于理解，专业性和应用性强。

本书可作为高职高专院校建筑类相关专业的英语教材，也可以作为其他专业学生的选修课教材及参考用书。

版权专有　侵权必究

图书在版编目(CIP)数据

建筑英语 / 郭二莹，姜华主编. —2版. —北京：北京理工大学出版社，2019.7（2024.6重印）
ISBN 978-7-5682-6871-4

Ⅰ.①建… Ⅱ.①郭…②姜… Ⅲ.①建筑—英语—高等学校—教材 Ⅳ.①TU-43

中国版本图书馆CIP数据核字（2019）第052162号

责任编辑：武丽娟	文案编辑：武丽娟
责任校对：杜　枝	责任印制：边心超

出版发行 / 北京理工大学出版社有限责任公司
社　　址 / 北京市丰台区四合庄路6号
邮　　编 / 100070
电　　话 / （010）68914026（教材售后服务热线）
　　　　　（010）68944437（课件资源服务热线）
网　　址 / http://www.bitpress.com.cn
版 印 次 / 2024年6月第2版第3次印刷
印　　刷 / 北京紫瑞利印刷有限公司
开　　本 / 787 mm×1092 mm　1/16
印　　张 / 15.5
字　　数 / 359千字
定　　价 / 55.00元

图书出现印装质量问题，请拨打售后服务热线，负责调换

第2版前言

"建筑英语"是高职高专院校建筑工程技术、建筑装饰工程技术、建筑工程管理等相关专业的必修课程。本书的编写力求满足该类专业学生对专门用途英语的需求，突出语言实用性及职业涉外交际能力，进一步提升学生职业素养和就业竞争力。

本书修订力求与时俱进，全面更新拓展阅读课文，利用二维码辅助教学。修订版参阅了国内外多部建筑书籍及期刊，将最新的建筑理念和建筑成果融入本书，使其更具有引导性和实效性。

本书的修订深入思考学生未来职业发展需求，本着实用为主，够用为度的原则，调整知识考核模式，增加了《中华人民共和国建筑法》和《建筑施工合同》中英文对照版，以满足课堂教学需求，同时提升学生职业素养。

本书共分为十个单元，涵盖了十个主题，主要内容包括世界建筑史、中国古代建筑、建筑设计、建筑材料、建筑结构、建筑施工、房屋设备、室内装饰、园林设计以及生态与建筑等。

本书由郭二莹、姜华担任主编，孙熙阳、刘玥、田妲妮担任副主编。

虽经反复讨论修改，但限于编者的学识及专业水平和实践经验，修订后的教材仍难免有疏漏和不妥之处，敬请广大读者指正。

编　者

第1版前言

作为专业英语课程，建筑英语是高职高专基础英语课程的后续课程，开设的目的是更好地满足高职高专建筑工程类专业学生对专业英语的就业需求，突出专业英语的实用性，培养借助于工具书阅读和翻译建筑类专业英语文章的能力以及简单的涉外交际能力，以此提高高职高专学生的职业素养，提升其未来就业竞争力。

本书是依据《高职高专教育英语课程教学基本要求（试行）》和《高等职业教育英语课程教学要求（试行）》，结合高职高专院校建筑工程类专业的课程设置和人才培养方案，并面向高职高专建筑工程类各专业的学生而编写的基础教材。本书建议授课学时为72学时，也可以根据实际情况调整。

本书共10个单元，每个单元都有一个主题，内容涉及世界建筑史、中国古代建筑、建筑设计、建筑材料、建筑结构、建筑施工、房屋设备、室内装饰、园林设计以及生态与建筑。每个单元由四部分组成：Text A为课内教学材料，文章后提供词汇及重点句式讲解，并配备针对性强的多种练习题以巩固教学重点、难点；Text B可作为课外拓展内容，更好地供学生自学，也提供专业词汇及难句的讲解；Oral Practices 为口语练习部分，紧密结合单元主题，注重学生实际交际能力的培养和训练；Translation Skills 旨在培养学生的专业英语翻译能力，遵照循序渐进的原则，从词汇、短语、从句等多方面讲解建筑英语的翻译技巧。

本书具有如下特点：

1. 内容丰富，选材广泛。文章选取涉及建筑专业活动的各个环节，同时涵盖了中外建筑史、著名建筑师以及绿色建筑等新的建筑理念；既介绍了中外古典建筑，也涉猎了大量最新的现代著名建筑，通过学习，学生在提高专业英语水平的同时还会扩大知识面。

2. 图文并茂，易于理解。基于高职学生相对薄弱的英语水平，书中的大量插图使教学内容更加直观，有利于激发学生的学习兴趣。

3. 专业性强。所选文章内容具有一定的专业性，口语练习围绕建筑工地施工进行，翻译技巧紧密结合本专业，能够多角度体现教学内容的专业性。

4. 应用性强。本书体系严密完整，弱化理论，强化实践内容，侧重技能传授。各章节内容均围绕培养学生的实际应用能力而展开。

本书既可以作为高职高专院校建筑类专业的专业英语教材，也可以作为其他各专业学生的选修课教材。

本书由郭二莹、崔雪、姜华任主编，孙熙阳、刘玥、田妲妮任副主编，高玉英、李云童、胡妍莉、张丹任参编。在编写过程中，我们参阅了大量的相关教材及网络文献，在此，对相关作者表示由衷感谢。同时，感谢付德成老师对本书的专业指导。

由于编者水平有限，书中难免有不妥之处，敬请读者批评指正，以便再版时改进和提高。

编　者

目 录

Unit One　History of World Architecture 世界建筑史 ·················· 1
　Text A　A Brief Introduction to the Evolution of World Architecture 世界建筑史简介 ······ 2
　Text B　New Seven Wonders of the World 世界"新七大奇迹" ······················ 6
　Oral Practices ·· 12
　Translation Skills—建筑英语翻译之词汇 ·· 13

Unit Two　Ancient Chinese Architecture 中国古代建筑 ···················· 16
　Text A　A Brief Introduction to the Ancient Chinese Architecture 中国古代建筑简介 ······ 17
　Text B　History of Chinese Architecture 中国建筑史 ······························ 21
　Oral Practices ·· 25
　Translation Skill—建筑英语翻译之语言结构特征 ·································· 26

Unit Three　Building Design 建筑设计 ···································· 29
　Text A　A Brief Introduction to the Building Design 建筑设计简介 ···················· 30
　Text B　Designers Redefining Modern Architecture 给现代建筑重新定义的建筑设计师 ········ 34
　Oral Practices ·· 38
　Translation Skill—建筑英语翻译之词义的选择及引申 ······························ 39

Unit Four　Building Materials 建筑材料 ·································· 41
　Text A　A Brief Introduction to the Building Materials 建筑材料简介 ·················· 42
　Text B　Little Ant Shadow Play Theater 小蚂蚁皮影剧场 ·························· 47
　Oral Practices ·· 51
　Translation Skill—建筑英语翻译之词义的增译和减译 ······························ 52

Unit Five　Building Structure 建筑结构 ·································· 55
　Text A　A Brief Introduction to the Building Structures 建筑结构简介 ·················· 56
　Text B　Cattle Back Mountain Volunteers' House 牛背山志愿者之家 ················· 60
　Oral Practices ·· 66
　Translation Skill—建筑英语翻译之定语从句的翻译 ································ 67

Unit Six　Building Construction 建筑施工 ································ 69
　Text A　A Brief Introduction to the Building Construction 建筑施工简介 ················ 70

  Text B Safety Factors in Design a Building 楼房设计中的安全因素 ······ 74
  Oral Practices ······ 79
  Translation Skills—建筑英语翻译之名词性从句的翻译 ······ 80

Unit Seven House Facilities 房屋设备 ······ 83
  Text A A Brief Introduction to the House Facilities 房屋设备简介 ······ 84
  Text B Tongling Recluse 铜陵山居 ······ 88
  Oral Practices ······ 94
  Translation Skills—建筑英语翻译之状语从句的翻译 ······ 95

Unit Eight Interior Decoration 室内装饰 ······ 98
  Text A Green Interior Decoration 绿色室内装饰 ······ 99
  Text B Five Elements of Interior Design in Decorating 室内装饰设计的三个基本因素 ······ 104
  Oral Practices ······ 107
  Translation Skills—建筑英语翻译之长句的翻译 ······ 108

Unit Nine Landscape Design 园林设计 ······ 110
  Text A Landscape Design Styles 景观设计风格 ······ 111
  Text B Famous Residential Landscape Design 著名的住宅景观设计 ······ 116
  Oral Practices ······ 121
  Translation Skills—建筑英语翻译之特殊句型的翻译 ······ 122

Unit Ten Ecology and Architecture 生态与建筑 ······ 126
  Text A Green Building 绿色建筑 ······ 127
  Text B Pushed Slab 推板办公楼 ······ 131
  Oral Practices ······ 136
  Translation Skills—建筑英语翻译之数量的翻译 ······ 137

Appendix I：课后习题答案 ······ 140

Appendix II：课文参考译文 ······ 143

Appendix III：词汇表 ······ 165

Appendix IV：建筑专业词汇中英文对照 ······ 186

Appendix V：建设工程施工合同 ······ 197

Appendix VI：中华人民共和国建筑法 ······ 218

参考文献 ······ 240

Unit One History of World Architecture
世界建筑史

Warming-up

Task 1: Match the English expressions with their corresponding equivalents.

1. the Sphinx   A. 帕特农神庙
2. Erechtheion   B. 十四圣徒朝圣教堂
3. Parthenon   C. 米拉之家
4. Church of Duomo   D. 水晶宫
5. Arch of Constantine   E. 伊瑞克提翁神庙
6. Basilica of the Fourteen Holy Helpers   F. 狮身人面像
7. the Crystal Palace   G. 米兰大教堂
8. CASA MILà   H. 君士坦丁凯旋门

Task 2: Match the pictures with the above famous architectures in Task 1.

1. _____

2. _____

3. _____

4. _____

5. _____

6. _____

7. _____

8. _____

Text A A Brief Introduction to the Evolution of World Architecture
世界建筑史简介

Early Architecture

Beginning with the first uses of brick and stone and ending with the completion of the great pyramids and **colossal Sphinx**, Ancient Egypt was home to some of the most influential architecture in history. The Nile Valley has been home to many of the richest civilizations in art, architecture, and design for at least ten thousand years, and this **innovation** began with a simple problem: a lack of wood[1].

The Egyptians were one of the first societies to seize upon the **durability** of bricks in construction, and their architectural monuments have endured thousands of years to become models of ancient architecture even in modern times[2]. Because of this durability, some of the most famous buildings in history were to come from the Nile Valley in Egypt.

Greek and Roman Architecture

As western society began to **bloom** and develop, architecture **took on** new life in the designs of the Greeks. One of the most powerful civilizations **rising up** in Europe, the Greek architects created history with their **flair** for order, design, and beauty[3]. The first standards of beauty, or the ideal **proportions**, were also Greek inventions; and every society following would **imitate** these ideals.

Two Greek architectural orders developed more or less concurrently. **The Doric order predominated** on the mainland and in the western colonies. **The Ionic order** originated in the cities on the islands and coasts of Asia Minor.

Rome would be the most major society to follow in Greek footsteps, creating some of the most famous buildings in the history of the world after the Grecian style. Rome became a powerful, well-organized **empire**, marked by great engineering works, roads, **canals**, bridges and **aqueducts**. Two Roman inventions allowed for greater architectural **flexibility**: the **dome** and the **groin**. The Romans also introduced the **commemorative**, **Triumphal Arch** and the **colosseum**, stadium.

Renaissance Architecture

Beginning in Italy in about 1400 A.D., Renaissance brought a revival of the **principles** and styles of ancient Greek and Roman architecture, it spreaded to the rest of Europe during the next 150 years. Perhaps architecture made its greatest leaps during the Renaissance. **Harmonious** form, **mathematical proportion** and artistic style combined to leave us such innovations as the stained glass window, the **Gothic cathedral**, the towering **spire**, and of course the **octagonal** dome[4].

Baroque and Rococo Architecture

After the Renaissance period, the architects of the era started to get bored with the **symmetry** and some old forms they had been using for the last 200 years. The most distinct shape of the Baroque style is the use of the **oval**. It appeared in many churches and was very characteristics of the time.

Rococo was the last phase of Baroque in France. It was a light-hearted, **decorative** style invented to suit the people of Paris. One of the main characteristic of the Rococo style was rooms that were meant to have music played in them.

The Industrial Architecture

The Industrial Revolution, which began in England in about 1760, brought a flood of new building materials, for example, **cast iron**, steel and glass[5]. Late 18th century designers and **patrons** turned towards the original Greek and Roman **prototypes**. In the 19th century, English architect Sir Joseph Paxton created the **Crystal Palace** (1850—1851) in London, a vast exhibition hall which **foreshadowed** industrialized buildings and the widespread use of cast iron and steel.

Modern Architecture

At the beginning of the 20th century, some designers refused to work in borrowed styles. Spanish architect **Antoni Gaudi** was the most original.

Between about 1965 and 1980, architects and critics began to support postmodernism. Although **postmodernism** is not a **cohesive** movement based on a distinct set of principles, in general postmodernists value individuality, **intimacy**, complexity, and occasionally humor.

By the early 1980s, postmodernism had become the dominant trend in American architecture and an important phenomenon in Europe as well.

New Words

colossal [kəˈlɒsl] *adj.* 巨大的，庞大的
Sphinx [sfɪŋks] *n.* [埃] 狮身人面
innovation [ˌɪnəˈveɪʃn] *n.* 改革，创新
durability [ˌdjʊərəˈbɪləti] *n.* 耐久性，持久性
bloom [bluːm] *n.* 最盛期，繁荣
flair [fleə(r)] *n.* 天资，天分
proportion [prəˈpɔːʃn] *n.* 比，比率
imitate [ˈɪmɪteɪt] *vt.* 模仿，效仿
predominate [prɪˈdɒmɪneɪt] *vi.* 占支配地位

empire [ˈempaɪə(r)] *n.* 帝国，帝国领土
canal [kəˈnæl] *n.* 运河
aqueduct [ˈækwɪdʌkt] *n.* 渡槽，引水渠
flexibility [ˌfleksəˈbɪləti] *n.* 机动性，灵活性
dome [dəʊm] *n.* 圆屋顶
groin [grɔɪn] *n.* 交叉拱
commemorative [kəˈmemərətɪv] *adj.* 纪念的，纪念性的
colosseum [ˌkɒləˈsiːəm] *n.* 竞技场，斗兽场

principle [ˈprɪnsəpl] n. 原则，原理，准则
harmonious [hɑːˈməʊniəs] adj. 和谐的，融洽的
cathedral [kəˈθiːdrəl] n. 大教堂
spire [ˈspaɪə(r)] n. 塔尖，尖顶
octagonal [ɒkˈtæɡənl] adj. 八边形的
symmetry [ˈsɪmətri] n. 对称，对称美
oval [ˈəʊvl] adj. 椭圆形的
decorative [ˈdekərətɪv] adj. 装饰的

patron [ˈpeɪtrən] n. 赞助人，资助人
prototype [ˈprəʊtətaɪp] n. 原型，雏形
foreshadow [fɔːˈʃædəʊ] vt. 预示，预兆
postmodernism [ˌpəʊstˈmɒdənɪzəm] n. 后现代主义
cohesive [kəʊˈhiːsɪv] adj. 有凝聚力的，紧密结合的
intimacy [ˈɪntɪməsi] n. 亲密，亲近

Phrases & Expressions

seize upon 利用
take on 呈现，具有
rise up 兴起

Triumphal Arch 凯旋门
mathematical proportion 数学比例
cast iron 铸铁

Proper Names

The Doric order 多立克柱式
The Ionic order 爱奥尼柱式
Renaissance [rɪˈneɪsns] 文艺复兴
Gothic [ˈɡɒθɪk] 哥特式
Baroque [bəˈrɒk] 巴洛克式

Rococo [rəˈkəʊkəʊ] 洛可可式
Joseph Paxton（1803—1865） 约瑟夫·帕克斯顿
Crystal Palace 水晶宫
Antoni Gaudi（1852—1926） 安东尼·高迪

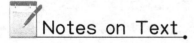

Notes on Text

[1] The Nile Valley has been home to many of the richest civilizations in art, architecture, and design for at least ten thousand years, and this innovation began with a simple problem: a lack of wood.

尼罗河流域在至少一万年间，一直都是艺术、建筑、设计这些灿烂文明的摇篮，这方面的创新均源于一个简单的原因：缺少木料。

[2] The Egyptians were one of the first societies to seize upon the durability of bricks in construction, and their architectural monuments have endured thousands of years to become models of ancient architecture even in modern times.

埃及人是最早利用经久耐用的砖结构的群体之一，他们的建筑古迹历经数千年，甚至从现代来看也是古代建筑的典范。

[3] As western society began to bloom and develop, architecture took on new life in the designs of the Greeks. One of the most powerful civilizations rising up in Europe, the Greek architects created history with their flair for order, design, and beauty.

随着西方社会的繁荣和发展，希腊人的建筑设计呈现新的活力。一个强大的文明在欧洲异军突起，希腊建筑师凭借他们在建筑法则、设计和美学方面的才华，在建筑史上留下了辉煌的一页。

[4] Harmonious form, mathematical proportion and artistic style combined to leave us such innovations as the stained glass window, the Gothic cathedral, the towering spire, and of course the octagonal dome.

和谐的外形、精确的数学比例、优美的格调结合在一起，让我们感受到许多创新之处，比如彩色玻璃窗、哥特式教堂、高耸的尖顶，当然还有八角形的穹顶。

[5] The Industrial Revolution, which began in England in about 1760, brought a flood of new building materials, for example, cast iron, steel and glass.

1760年始于英国的工业革命让很多新的建筑材料涌现出来，比如铸铁、钢和玻璃。

Exercises

Ⅰ. Answer the following questions according to the text.

1. What is the main reason for the architectural innovation of the Nile Valley?
2. What are the main marks of Roman as a powerful, well-organized empire?
3. Why did Renaissance bring a revival of ancient Greek and Roman architecture?
4. What foreshadows industrialized buildings and the widespread use of cast iron and steel?
5. When did postmodernism become the dominant trend in American architecture?

Ⅱ. Match the English words with their Chinese equivalents.

1. innovation   A. 比，比率
2. aqueduct   B. 原型，雏形
3. dome   C. 交叉拱
4. groin   D. 原则，原理
5. colosseum   E. 八边形的
6. principle   F. 对称，对称美
7. octagonal   G. 渡槽，引水渠
8. symmetry   H. 竞技场，斗兽场
9. prototype   I. 改革，创新
10. proportion   J. 圆屋顶

Ⅲ. Read the following passage and fill in the blanks with the words given in the box.

| predominant | erected | limestone | fertilizer | reserved |
| molds | subsidiary | harden | granite | inaccessible |

Due to the scarcity（缺乏）of wood, the two _____ building materials used in ancient Egypt were sun-baked mud brick and stone, mainly _____, but also sandstone and _____ in considerable quantities. From the Old Kingdom onward, stone was generally _____ for tombs and temples, while bricks were used even for royal palaces, fortresses（要塞）, the walls of temple precincts（内部）and towns, and for _____ buildings in temple complexes.

Ancient Egyptian houses were made out of mud collected from the Nile River. It was placed in _____ and left to dry in the hot sun to _____ for use in construction.

Many Egyptian towns have disappeared because they were situated near the cultivated area of the Nile Valley and were flooded as the river bed slowly rose during the millennia（一千年）, or the mud bricks of which they were built were used by peasants as _____. Others are _____, new buildings having been _____ on ancient ones. Fortunately, the dry, hot climate of Egypt preserved some mud brick structures. Also, many temples and tombs have survived because they were built on high ground unaffected by the Nile flood and were constructed of stone.

Ⅳ. Translate the following English into Chinese or Chinese into English.

1. the Nile Valley
2. Renaissance Architecture
3. building material
4. mathematical proportion
5. Triumphal Arch
6. 多立克柱式
7. 巴洛克式
8. 哥特式教堂
9. 后现代主义
10. 主导趋势

Text B New Seven Wonders of the World
世界"新七大奇迹"

Inspired by the Seven Wonders of Ancient World, an unofficial Organization decides to crown seven new world wonders among the **candidates**. July 7th, 2007 the new seven world wonders were nominated in Lisbon, Portugal. Gloriously crowned, the following were chosen as the New Seven Wonders of the World from among the 21 candidates by the people all over the world.

The Great Wall, China (220 B.C., 1368 A.D.—1644 A.D.)

The Great Wall of China was built to link existing **fortifications** into a united defense system and better keep invading Mongol tribes out of China[1]. It is the largest man-made monument ever to have been built and it is disputed that it is the only one visible from space. Many thousands of people must have given their lives to build this colossal construction (Fig.1-1).

Fig.1-1　The Great Wall, China

Petra, Jordan (9 B.C.—40 A.D.)

On the edge of the Arabian Desert, Petra was the glittering capital of the Nabataean empire of King Aretas IV (9 B.C. to 40 A.D.). Masters of water technology, the Nabataeans provided their city with great **tunnel** constructions and **water chambers**. A theater, modeled on Greek-Roman prototypes, had space for an audience of 4,000. Today, the Palace Tombs of Petra, with the 42-meter-high **Hellenistic** temple facade on the El-Deir **Monastery**, are impressive examples of Middle Eastern culture[2] (Fig.1-2).

Fig.1-2　Petra, Jordan

The Roman Colloseum, Italy (70 A.D.—82 A.D.)

This great **amphitheater** in the centre of Rome was built to give favors to successful **legionnaires** and to celebrate the glory of the Roman Empire. Its design concept still **stands to** this very day, and **virtually** every modern sports stadium some 2,000 years later still bears the **irresistible** imprint of the Colosseum's original design[3]. Today, through films and history books, we are even more aware of the cruel fights and games that **took place** in this **arena**, all for the joy of the spectators (Fig.1-3).

Fig.1-3　The Roman Colloseum, Italy

Machu Picchu, Peru (1460 A.D.—1470 A.D.)

In the 15th century, the Incan Emperor Pachacútec built a city in the clouds on the mountain known as Machu Picchu ("Old Mountain") (Fig.1-4). This extraordinary settlement lies halfway up the Andes

Fig.1-4　Machu Picchu, Peru

Plateau, deep in the Amazon jungle and above the Urubamba River[4]. It was probably abandoned by the Incas because of a **smallpox** outbreak and, after the Spanish defeated the Incan Empire, the city remained "lost" for over three centuries. It was rediscovered by Hiram Bingham in 1911.

Chichén Itzá, Mexico (Before 800 A.D.)

Fig.1-5　Chichén Itzá, Mexico

Chichén Itzá, the most famous **Mayan** temple city, served as the political and economic center of the Mayan civilization. Its various structures—the Pyramid of Kukulkan, the Temple of Chac Mool, the Hall of the Thousand Pillars, and the Playing Field of the Prisoners—can still be seen today and are **demonstrative** of an extraordinary commitment to architectural space and composition[5]. The pyramid itself was the last, and arguably the greatest, of all Mayan temples (Fig.1-5).

The Taj Mahal, India (About 1631 A.D.— 1653 A.D.)

Fig.1-6　The Taj Mahal, India

This immense **mausoleum** was built on the orders of Shah Jahan, the fifth Muslim Mogul emperor, to honor the memory of his beloved late wife. Built out of white marble and standing in formally laid-out walled gardens, the Taj Mahal **is regarded** as the most perfect jewel of Muslim art in India[6]. The emperor was consequently jailed and, it is said, could then only see the Taj Mahal out of his small cell window (Fig.1-6).

Christ Redeemer, Brazil (1931 A.D.)

Fig.1-7　Christ Redeemer, Brazil

This statue of Jesus stands some 38 meters tall, **atop** the Corcovado Mountain overlooking Rio de Janeiro. Designed by Brazilian Heitor da Silva Costa and created by French sculptor Paul Landowski, it is one of the world's best-known monuments. The statue took five years to construct and was **inaugurated** on October 12, 1931. It has become a symbol of the city and of the warmth of the Brazilian people, who receive visitors with open arms (Fig.1-7).

New Words

inspire [ɪnˈspaɪə(r)] vt. 鼓舞，激励
candidate [ˈkændɪdət] n. 候选，候选人
fortification [ˌfɔːtɪfɪˈkeɪʃn] n. 筑垒，防御工事
monument [ˈmɒnjumənt] n. 纪念碑，遗迹，遗址
glittering [ˈɡlɪtərɪŋ] adj. 辉煌的，光辉灿烂的
tunnel [ˈtʌnl] n. 隧道，地道
Hellenistic [ˌheliˈnɪstɪk] adj. 希腊风格的，希腊文化的
monastery [ˈmɒnəstri] n. 修道院，寺院
amphitheater [ˈæmfɪθɪətə(r)] n. 竞技场，斗兽场
legionnaire [ˌliːdʒəˈneə(r)] n. 军团的士兵

virtually [ˈvɜːtʃuəli] adv. 实际上，事实上
irresistible [ˌɪrɪˈzɪstəbl] adj. 无法抗拒的，不可阻挡的
arena [əˈriːnə] n. 舞台，竞技场
smallpox [ˈsmɔːlpɒks] n. 天花
Mayan [ˈmɑːjən] adj. 玛雅人的，玛雅语的；n. 玛雅人，玛雅语
demonstrative [dɪˈmɒnstrətɪv] adj. 说明的，表明的
mausoleum [ˌmɔːsəˈliːəm] n. 陵墓
atop [əˈtɒp] adv. & prep. 在……顶上
inaugurate [ɪˈnɔːɡjəreɪt] vt. 开创，举行典礼

Phrases & Expressions

water chambers 水箱，蓄水池
stand to 坚持，不放弃

take place 发生，进行
be regarded as 视为，被认为是

Proper Names

Petra 佩特拉古城（约旦）
Arabian Desert 阿拉伯沙漠
Nabataean 纳巴泰人
Pachacútec 帕查库蒂（印加统治者）
Machu Picchu 马丘比丘（秘鲁）
Andes 安第斯山脉
Urubamba River 乌鲁班巴河谷
Amazon jungle 亚马孙热带丛林
Incan Empire 印加帝国

Hiram Bingham 海勒姆·宾厄姆（耶鲁大学考古学家）
Chichén Itzá 奇琴伊察金字塔
Pyramid of Kukulkan 库库尔坎金字塔
Temple of Chac Mool 查尔穆尔神殿
Hall of the Thousand Pillars 千柱林
Corcovado Mountain 科尔科瓦多山
Rio de Janeiro 里约热内卢（巴西城市）

Notes on Text

[1] The Great Wall of China was built to link existing fortifications into a united defense system and better keep invading Mongol tribes out of China.

中国的万里长城是中国古代为抵御蒙古部落入侵而建造的，它将已有的单个要塞连成一体，从而形成一个完整的防御体系。

[2] Today, the Palace Tombs of Petra, with the 42-meter-high Hellenistic temple facade on the El-Deir Monastery, are impressive examples of Middle Eastern culture.

今天，在埃尔代尔的修道院，佩特拉陵寝与42米高的希腊寺庙面对面矗立，这已成为中东文化遗产的重要代表。

[3] Its design concept still stands to this very day, and virtually every modern sports stadium some 2,000 years later still bears the irresistible imprint of the Colosseum's original design.

斗兽场的建筑设计并不落后于现代的美学观点，而事实上，大约2 000年后的今天，每一个现代化的大型体育场都或多或少地烙上了一些古罗马斗兽场的设计风格。

[4] This extraordinary settlement lies halfway up the Andes Plateau, deep in the Amazon jungle and above the Urubamba River.

这个非凡的居住地上延至安第斯山脉，下潜至亚马孙热带丛林，并位于乌鲁班巴河谷之上。

[5] Its various structures—the pyramid of Kukulkan, the Temple of Chac Mool, the Hall of the Thousand Pillars, and the Playing Field of the Prisoners-can still be seen today and are demonstrative of an extraordinary commitment to architectural space and composition.

城内至今仍可见的古迹主要有库库尔坎金字塔、查尔穆尔神殿、千柱林和囚犯竞技场。这些建筑在空间和造型组合上均充分体现出了玛雅人杰出灵动的建筑意识。

[6] Built out of white marble and standing in formally laid-out walled gardens, the Taj Mahal is regarded as the most perfect jewel of Muslim art in India.

陵墓由白色大理石砌成，坐落于有高大围墙的花园中，非常宏伟壮观。泰姬陵被公认为穆斯林建筑艺术在印度最杰出、最完美的代表。

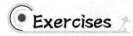

Exercises

I. Choose the best answers according to the text.

1. Where are the new seven world wonders nominated _____ on July 7th, 2007?
   A. in Madrid, Spain   B. in Rome, Italy
   C. in Paris, France   D. in Lisbon, Portugal

2. What is the purpose of building the Great Wall of China?
   A. To invade Mongol tribes.
   B. To prevent the invasion of Mongol tribes.
   C. To keep the unity of China.
   D. To link two tribes.

3. What are Nabataeans skillful at?
   A. Civil engineering technology.   B. Water technology.
   C. Building technology.   D. Digital technology.

4. Why was Machu Picchu probably abandoned by the Incas?
   A. Because a smallpox broke out.   B. Because a virus broke out.
   C. Because an influenza broke out. D. Because an insect broke out.

5. As the political and economic center of the Mayan civilization, there are various structures in Chichén Itzá except _____.
   A. Temple of Chac Mool   B. Hall of the Thousand Pillars
   C. Pyramid of Kukulkan   D. Temple of Heaven

6. Christ Redeemer has become a symbol of _____.
   A. Mexico   B. Argentina
   C. Brazil   D. Chili

II. Fill in the blanks with the following words. Change the forms if necessary.

| inspire | candidate | monument | tunnel |
| virtually | arena | irresistible | inaugurated |

1. Mr. Putin was _____ as the President of the Russian Federation.
2. The athletes marched into the _____ to the sound of music.
3. Rescuers used props to stop the roof of the _____ collapsing.
4. We have adopted Mr. Stevens as our _____ at the next election.
5. A good novel serves to educate and _____ people.
6. The _____ was erected to the memory of the people's heroes.

7. This new material is _____ unbreakable.

8. He felt an _____ impulse to rush into the room.

III. Translate the following English into Chinese or Chinese into English.

1. defense system
2. water chambers
3. Mayan temple
4. cell window
5. Hall of the Thousand Pillars
6. 安第斯山脉
7. 亚马孙热带丛林
8. 查尔穆尔神殿
9. 基督像
10. 民间组织

Oral Practices

Conversation One

A: Good morning! My name is Peter. Nice to meet you!

B: Good morning! I'm John. Nice to meet you!

A: What do you major in, Peter?

B: I major in Architectural Engineering.

A: What are the main courses of your major this term?

B: They are History of World Architecture, Architectural Design, Building Mechanics and Architectural Economics.

A: I am interested in the medieval European architecture. Can you introduce something for me briefly?

B: Sure. In the medieval period, many different styles of architecture developed. For example, churches that tried to imitate Roman vaults and arches are now called "Romanesque." Later churches that used pointed arches and more decorative window tracery are called "Gothic", which is subdivided into different local styles.

A: Really? It's so interesting. It's time for lunch. Let's go together and share the opinions with each other.

B: That's a good idea.

Conversation Two

A: Welcome home. How was your trip to South America? Did you see some amazing architecture?

B: It was excellent. And yes, I saw a great deal of architecture. We spent most of our time on

Mayan territory.

A: I've heard you will get the flavor of the place as soon as you get there.

B: The pyramids and temples were predominant. The stone work was absolutely superb.

A: What I found most interesting was the way they were able to make them appear rustic and yet exquisite at the same time. The pyramids were remarkably compact and predominantly made of stone.

B: I hope you take the time to reflect on your visit there and really value the experience. It's not often that you get a chance to visit such an elaborate and ethnic site. I'm glad you went.

Translation Skills——建筑英语翻译之词汇

科技英语是从事科学技术活动时所使用的英语，是英语的一种变体（科技文体）。建筑英语是科技英语的分支之一，其在翻译中注重科学性、逻辑性、正确性与严密性，在词汇、语法、修饰等方面具有其自身的特点。建筑英语词汇翻译成汉语时，不仅要注意单词词义的选择，力求"信"，同时要合乎专业规范术语的要求，为句子翻译的"达"打下基础。下面将着重介绍建筑英语词汇的主要构成形式及词汇的词性转译。

1. 词汇的构成形式

（1）合成法：将两个或两个以上的词组合成一个新词。

【例】waterproof 防水的　　floor-space 建筑面积　　item-test 项目检验

（2）拼缀法：是指在一个旧词前或后加上词缀构成新词的方法。

【例】anti-sweat 防结露　　pre-qualification 预审　　semiconductor 半导体

（3）混成法：将原有二词各取其中一部分（有时还是某一词的全部）合成新词。

【例】escalator + lift → escalift 自动电梯

（4）缩略法：以首字母缩略为主，即将某一词组中的几个主要词的首字母合起来组成新词。

【例】EC = engineering change 工程更改

　　　EDP = engineering design plan 工程设计方案

2. 词汇的词性转译

（1）转译成动词。

①介词转译成动词。

【例】Rigid pavements are made *from* rigid cement concrete.

刚性路面是用刚性水泥混凝土建造的。

【例】The Eads bridge was completed in 1874 over the Mississippi *at* St Louis, Missouri.

伊兹桥建成于1874年，它位于密苏里州的圣路易斯，横跨密西西比河。

②名词转译成动词。

【例】Global warming observed over the past century has focused attention on *mantled* greenhouse gas, presumed to be the likely cause.

在过去的一个世纪中观察到的全球变暖趋势，据推测可能是由人为的温室气体造成的。

【例】The most important point in building design is the *layout* of rooms, which shall provide the great possible convenience in relation to the purposes for which they are intended.

房屋具体设计的重点就是布置房间，它将为与使用要求相关的目的提供最大可能的便利。

③形容词转译成动词。

【例】They are *confident* that they will be able to build the modern bridge in a short time.

他们确信在短时间内能够建成这座现代化桥梁。

【例】Copper wire is *flexible*.

铜线容易弯曲。

④副词转译成动词。

【例】The construction of the dam was two months *behind*.

这座大坝的施工工期延误了两个月。

【例】The test of the new green building material was not *over* yet.

新型绿色建筑材料的试验还没结束。

（2）转译成名词。

①动词转译成名词。

【例】Concrete is *characterized* by its strength.

混凝土的特点是它的抗压强度。

②形容词转译成名词。

【例】A more *efficient* type of building is the apartment house.

公寓住宅是利用率较高的建筑类型。

【例】The more carbon the steel contains, the *harder* and *stronger* it is.

钢含碳量越高，强度和硬度就越大。

③副词转译成名词。

【例】The design must be *dimensionally* correct.

设计的尺寸必须正确。

【例】The location of the tunnel is shown *schematically* on this page.

这一页上所表示的是隧道位置的简图。

（3）转译成形容词。

①名词转译成形容词。

【例】The methods of prestressing a structure show considerable *variety*.

对结构施加预应力的方法是多种多样的。

【例】In this case to reinforce the joists to help them support the additional loads is an absolute *necessity*.

在这种情况下加固托梁有助于支撑附加荷载是绝对必要的。

②副词转译成形容词。

【例】It is a fact that no structural material is *perfectly* elastic.

事实上，没有一种结构材料是完全的弹性体。

【例】The landscape engineer had prepared *meticulously* for his design.

园林工程师为这次设计作了十分周密的准备。

（4）转译成副词。

①形容词转译成副词。

【例】Engineers have made a *careful* study of the properties of these new structures.

工程师们仔细地研究了这些新型结构的特性。

【例】With *slight* modifications each type can be used for all three systems.

每种型号只要稍加改动就能用于这三种系统。

②名词转译成副词。

【例】We find *difficulty* in solving this problem.

我们觉得难以解决这个问题。

【例】The added device will ensure *accessibility* for part loading and unloading.

增添这种装置将保证工件装卸方便。

Unit Two Ancient Chinese Architecture
中国古代建筑

Warming-up

Task 1: Match the English expressions with their corresponding equivalents.

1. Tiananmen   A. 长城
2. Potala Palace   B. 天安门
3. The Great Wall   C. 布达拉宫
4. Qianxun Pagoda   D. 石拱桥
5. Quadrangle Courtyard E. 大昭寺
6. Taihe Palace   F. 四合院
7. Chinese Stone Bridge G. 千寻塔
8. Jokhang Temple   H. 太和殿

Task 2: Match the pictures with the above famous architectures in Task 1.

1. _____

2. _____

3. _____

4. _____

5. _____

6. _____

7. _____

8. _____

Text A   A Brief Introduction to the Ancient Chinese Architecture
中国古代建筑简介

As an ancient **civilized** nation and a great country on the East Asian **continent**, China possesses a vast **territory** covering 9.6 million sq. km. and a population **accounting for** nearly one-fifth of the world's total, 56 nationalities and a history of 5,000 years, during which it has created a unique, outstanding traditional Chinese architecture that is a particularly beautiful branch in the tree of Chinese **civilization**[1].

Architecture Design Ideas

All ancient Chinese architecture was built **according to** strict rules of designing that made Chinese building follow the ideas of **Taoism** or other Chinese **philosophies**[2]. The first designing idea was that buildings should be long and low rather than short and tall—they should seem almost to be hugging you. The roof would be held up by **columns**, and not by the walls. The roof should seem to be floating over the ground. The second designing idea was symmetry: both sides of the building should be the same, balanced, just as Taoism **emphasized** balance (Fig.2-1). Even as early as the **Shang Dynasty**, about 1500 B.C., Chinese buildings looked pretty much like this, with **curved tile** roofs and long rows of **pillars**.

Fig.2-1   Pavilion of Prince Teng

Architecture Features

Chinese ancient architecture **constitutes** the only system based mainly on wooden structures of unique charming appearance. This differs from all other **architectural** systems in the world which are based mainly on **brick** and stone structures. Wooden **posts**, **beams**, **lintels** and **joists** make up the framework of a house. Walls serve as the separation of rooms without bearing the weight of the whole house, which is unique to China[3]. As a famous saying goes, "Chinese house will still stand when their walls **collapse**." The **specialty** of wood requires **antisepsis** methods to be **adopted**, thus develops into Chinese own architectural painting decoration[4]. Colored **glaze** roofs, windows with **exquisite appliqué** design and beautiful flower patterns on wooden pillars reflect the high-level of the craftsmen's handicraft and their rich imagination.

Architecture and Culture

Architecture and culture are tightly related to each other. **In a sense**, architecture is the carrier of culture. Styles of Chinese ancient architecture are rich and varied, such as temples, **imperial** places, **altars**, **pavilions**, official residences and folk house, which greatly **reflect** Chinese ancient thoughts—the harmonious unity of human beings with nature (Fig.2-2).

There are two typical types of the Chinese ancient architecture representing the profound influence of Chinese culture, which are Feng Shui and Memorial Arch.

Feng Shui: Chinese traditional theory especially directs the process of architectural construction **on the basis of** the culture of **The Book of Changes**. Its emphasis is concerned with the harmonious unity of human beings with nature.

Memorial Arch: It is the derivative of Chinese feudal society, also called Pailou, unique to China, was built to honor great achievement and virtue of ancestors[5] (Fig.2-3).

Today based on its traditional soil, Chinese architecture has absorbed foreign architectural culture and continued to **forge ahead** by **complying with** the requirements of our time and using new architectural techniques.

Fig.2-2　LongShan Temple

Fig.2-3　Stone Arches

New Words

rchitecture [ˈɑːrkɪtektʃə(r)] n. 建筑，建筑学
civilize [ˈsɪvəlaɪz] v. 使文明，教化
continent [ˈkɒntɪnənt] n. 洲，大陆
territory [ˈterətri] n. 领土，范围
civilization [ˌsɪvəlaɪˈzeɪʃn] n. 文明，文化
philosophy [fəˈlɒsəfi] n. 哲学
column [ˈkɒləm] n. 柱，圆柱
emphasize [ˈemfəsaɪz] v. 强调
dynasty [ˈdɪnəsti] n. 王朝，朝代
curved [kɜːvd] adj. 弧形的，曲线的
tile [taɪl] n. 瓦，瓷砖
pillar [ˈpɪlə(r)] n. 柱子，支柱
feature [ˈfiːtʃə(r)] n. 特征，特点
constitute [ˈkɒnstɪtjuːt] v. 构成，组成

architectural [ˌɑːkɪˈtektʃərəl] adj. 建筑的，建筑学的
brick [brɪk] n. 砖
post [pəʊst] n. 柱，桩，杆
beam [biːm] n. 梁
lintel [ˈlɪntl] n. 过梁，（门或窗的）楣
joist [dʒɔɪst] n. 托梁，搁栅
collapse [kəˈlæps] vt. 使倒塌，使坍塌
specialty [ˈspeʃəlti] n. 独特之处，特点
adopt [əˈdɒpt] vt. 采用，采取
antisepsis [ˌæntɪˈsepsɪs] n. 防腐，消毒
glaze [gleɪz] v. 给……上釉，使光滑，使光亮
exquisite [ɪkˈskwɪzɪt] adj. 精致的，优美的
appliqué [əˈpliːkeɪ] n. 贴花，嵌花，补花

imperial [ɪmˈpɪərɪəl] *adj.* 帝国的，皇帝的
altar [ˈɔːltə(r)] *n.* 祭坛，圣坛
pavilion [pəˈvɪlɪən] *n.* 亭，阁楼
reflect [rɪˈflekt] *vt.* 反映

profound [prəˈfaʊnd] *adj.* 深厚的，意义深远的
absorb [əbˈzɔːb] *v.* 吸收

Phrases & Expressions

account for 说明，占，解决，得分
according to 依照
rather than 与其……倒不如……
in a sense 在某种意义上

on the basis of 以……为基础
Memorial Arch（Paifang）牌坊
forge ahead 稳步前进，开拓进取
comply with 遵从，服从

Proper Names

Taoism 道教
Shang Dynasty 商朝

The Book of Changes 《易经》

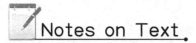

Notes on Text

[1] ...during which it has created a unique, outstanding traditional Chinese architecture that is a particularly beautiful branch in the tree of Chinese civilization.

中国悠久的历史创造了灿烂的古代文明，而独特杰出的中国传统建筑便是其重要的组成部分。

[2] All ancient Chinese architecture was built according to strict rules of designing that made Chinese building follow the ideas of Taoism or other Chinese philosophies.

所有中国古代建筑都是根据严格的设计规范来建造的，其一直追随道教和其他各学派的传统理念来设计。

[3] Walls serve as the separation of rooms without bearing the weight of the whole house, which is unique to China.

墙壁起分隔房间的作用而不是支撑作用，这是中国古代建筑的独特之处。

[4] The specialty of wood requires antisepsis methods to be adopted, thus develops into Chinese own architectural painting decoration.

因为木料耐久性差，需要采取专业的防腐措施，后来就发展成中国自己的建筑彩绘艺术。

· 19 ·

[5] Memorial Arch: It is the derivative of Chinese feudal society, also called Pailou, unique to China, was built to honor great achievement and virtue of ancestors.

牌坊：也叫作牌楼，是中国独一无二的封建社会文化的产物，是封建社会为表彰功勋和忠孝节义所建立的建筑物。

Exercises

Ⅰ. **Answer the following questions according to the text.**

1. What are the main designing ideas of ancient Chinese architecture?
2. What's the difference between Chinese architecture and other architectural systems?
3. What can reflect the high-level of the craftsmen's handicraft of ancient Chinese architecture?
4. What's the relationship between architecture and culture?
5. What is the significance of Feng Shui and Memorial Arch?

Ⅱ. **Match the English words with their Chinese equivalents.**

1. architecture   A. 哲学
2. curved   B. 强调
3. civilization   C. 吸收
4. philosophy   D. 建筑，建筑学
5. emphasize   E. 弧形的，曲线的
6. profound   F. 帝国的，皇帝的
7. absorb   G. 文明
8. imperial   H. 深厚的，意义深远的
9. brick   I. 领土，范围
10. territory   J. 砖

Ⅲ. **Read the following passage and fill in the blanks with the words given in the box.**

| sections | stretching | capital | lay | stately |
| south | palace | emperor | buildings | located |

The Forbidden City was built from 1406 to 1420 by the third Ming Emperor Zhu Di, who determined to move his _____ north from Nanjing to Beijing. It lost its original function in the year of 1924 when Pu Yi, the last _____ of the Qing Dynasty, was expelled（驱逐）from this palace after his abdication（退位）. During its 500-year span, twenty-four emperors lived and ruled from this palace.

The Forbidden City, _____ in the center of the axial（轴）line, stretches 763 meters from east to west and 961 meters from north to _____, covering 720,000 square meters（74 hectares）. There are 800 _____ that have in total about 9,000 rooms. The Forbidden City is the world's largest _____ complex. In line with the total length of the axial line _____ from south to north, the composition of the entire city can be divided into three major _____. The first section started from Yongdingmen to Zhengyangmen in the middle of the southern wall of the outer city. It was the longest, while its rhythm was the gentlest. The second section, where _____ the climax（高点）, was shorter and the most _____, running from Zhengyangmen to Jingshan through the square in front of the palace and the entire palatial（宫殿式的）city. The third section running from Jingshan to the Bell and Drum Towers was the shortest.

IV. Translate the following English into Chinese or Chinese into English.

1. East Asian continent
2. framework of a house
3. antisepsis methods
4. official residences
5. feudal society
6. 丰富的想象力
7. 传统设计
8. 深远影响
9. 结构体系
10. 宫殿

Text B History of Chinese Architecture
中国建筑史

Chinese architecture is as old as Chinese civilization, of which it is an **integral** part[1]. Considering social, economical, political and technological developments reflected in architecture, the developments of Chinese architecture can be **divided into** the following six main stages.

The First Stage: Before 220 A.D.

Starting from the early primitive society, to the **decline** of the Han Dynasty in 220 A.D., this stage marks the Chinese architectural development from start-up to early mature stage. Natural cave is a natural way to solve the housing problem of our ancestors.

From 206 B.C. to 220 A.D., China entered into the stage of fully matured **feudalism**. The palaces of the Han **emperors** were so magnificent that they had "a thousand doors and ten thousand windows." To win the blessings of gods, very high towers were built in the royal garden.

Meanwhile, business with West Asia brought in the West cultural factors in architecture, such as **arch**, human figures used as architectural **embellishments**, etc. Introduction of **Buddhism** also greatly **enriched** the architecture of China.

The Second Stage: Buddhist Influence, 3rd to 6th Century

In this period, Chinese architecture was greatly influenced by, and absorbed most from, Indian and **Central Asian culture**. Buddhist temples and pagodas were built almost everywhere.

The principal remains of this period are many rock-cut or cave temples and some rare masonry **pagodas**. Some of the cave temples, such as those at Yungang, Maiji Shan and Tianlong are architected in high standard. The oldest Buddhist pagoda in China today was also built in this period. Henan Songshan Song Yue Temple, built in A.D. 520, about 40 meters high and embellished with a few **motifs** of Indian origin, symbolized a new architectural form in China[2].

The Third Stage: the Glorious Tang Dynasty, 7th to 9th Century

By the end of the sixth century, China was unified again. Paintings, **sculpture**, music, dancing were all enriched with Western elements. Architecture also reached its heyday both technologically and artistically. The oldest specimen of the timber structures existing in China **dated back** to Tang Dynasty, which **illustrated** the carpenter's mastery art[3]. The wooden pagoda was gradually substituted by masonry structures. New forms of pagodas were innovated constantly. "**Multi-storied pagoda**" and "**multi-eaved pagoda**" appeared in this period.

The Fourth Stage: Large Ensembles Standardization and Refinement, 10th to 14th Century

The main developments of this stage are the expenses in the architectural groups' scales and the advances made in architectural refinements. In order to save materials and labor, speed the design and construction process of large-scale buildings, the method of stylization and standardization was employed[4]. The compilation and publication of the Ying-tsao-fa-shih is an historical event in the development of Chinese architecture.

There are still a considerable number of **edifices** from this period standing today. Except for few bridges, all are religious buildings—temples of timber construction and pagodas in masonry structures. On the pagoda layout, stark cross **axis** layout is introduced. Pagodas of this period are built in various forms. Instead of the square layout, the **octagon** became the most popular one then and thereafter.

The Fifth Stage: Age of Peking and European Influence, 1368—1949

Innumerable buildings in Ming and Qing Dynasties survived from the twice expansion of Peking. The city, with its palaces and gardens, is the most important historical sites and one of the most **magnificent** architectural works.

During the nineteenth century, capitalism and **imperialism** brought Western architectural concepts to China. In some coastal cities, even the city of the nineteenth-century Europe was reproduced in the "concessions" acquired by imperialists[5].

The architecture of the first half of the twentieth century in China reflects most faithfully the features of semi-feudalistic and semi-colonial Chinese society. In the countryside and smaller towns, traditional Chinese building is still people's choice. But in larger cities, the **"foreign-style house" prevails**.

The Sixth Stage: the Age of People's Architecture, after 1949

In 1949, the Chinese people liberated themselves and founded the People's Republic of China. A new epoch had begun for Chinese architecture.

New Words

integral [ˈɪntɪɡrəl] adj. 完整的
decline [dɪˈklaɪn] n. 下降，减少
feudalism [ˈfju:dəlɪzəm] n. 封建制度
emperor [ˈempərə(r)] n. 皇帝，君主
arch [ɑ:tʃ] n. 弓形（物）
embellishment [ɪmˈbelɪʃmənt] n. 装饰，修饰
Buddhism [ˈbʊdɪzəm] n. 佛教
enrich [ɪnˈrɪtʃ] vt. 使富裕，使富有
pagodas [pəˈɡəʊdə] n. 宝塔
motif [məʊˈti:f] n. 装饰图案

sculpture [ˈskʌlptʃə(r)] n. 雕刻
illustrate [ˈɪləstreɪt] vi.（用图、实例等）说明，阐明（+with）
edifices [ˈedɪfɪs] n. 大建筑物
axis [ˈæksɪs] n. 轴
octagon [ˈɒktəɡən] n. 八边形
innumerable [ɪˈnju:mərəbl] adj. 无数的
magnificent [mæɡˈnɪfɪsnt] adj. 宏伟的
imperialism [ɪmˈpɪərɪəlɪzəm] n. 帝国主义
prevail [prɪˈveɪl] v. 流行，战胜

Phrases & Expressions

be divided into 被分解
Central Asian culture 中亚文化
date back to 追溯至

multi-storied pagoda 楼阁式塔
multi-eaved pagoda 多重檐佛塔
foreign-style house 洋房

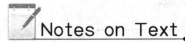

Notes on Text

[1] Chinese architecture is as old as Chinese civilization, of which it is an integral part.
中国建筑就像中国文明一样古老，可谓自成一体。

[2] Henan Songshan Song Yue Temple, built in A.D. 520, about 40 meters high and embellished with a few motifs of Indian origin, symbolized a new architectural form in China.

河南省嵩山嵩岳寺塔，建于公元520年，高约40米，雕饰有一些印度图案，标志着一种新的建筑形式在中国的出现。

[3] The oldest specimen of the timber structures existing in China dated back to Tang Dynasty, which illustrated the carpenter's mastery art.

中国现存最早的木结构建筑典范就出自唐代，从中可见木匠工艺的精湛程度。

[4] In order to save materials and labor, speed the design and construction process of large-scale buildings, the method of stylization and standardization was employed.

为节约材料和劳力，并且加快大规模建筑物的设计速度和建造进程，出现了对典型化和标准化的要求。

[5] In some coastal cities, even the city of the nineteenth-century Europe was reproduced in the "concessions" acquired by imperialists.

一些中国沿海城市，在帝国主义者得到特许的"租界"中，甚至复制了不少19世纪欧洲的城市建筑。

Exercises

I. Choose the best answers according to the text.

1. How many stages of the development of Chinese architecture are divided into?
   A. Five.　　　　　B. Six.　　　　　C. Seven.　　　　　D. Eight.

2. Why were many high towers built in the palaces before 220 A.D.?
   A. Because the Han emperors wanted to show their power.
   B. Because the Han civilians wanted to get the wishes of gods.
   C. Because the Han emperors wanted to get the blessing of gods.
   D. Because the Han emperors wanted to bless gods.

3. Which stage was affected greatly by Indian and Central Asian culture?
   A. The first stage.　　　　　　　B. The second stage.
   C. The third stage.　　　　　　　D. The fourth stage.

4. Chinese architecture reached the most powerful stage in the _____.
   A. Han Dynasty　　　　　　　　B. Tang Dynasty
   C. Ming Dynasty　　　　　　　　D. Song Dynasty

5. In the 19th century, _____ influenced Chinese architecture deeply.
   A. Eastern architectural concepts　　　B. Italian architectural concepts
   C. American architectural concepts　　D. Western architectural concepts

II. **Fill in the blanks with the following words. Change the forms if necessary.**

integral   magnificent   enrich   miracle
illustrate   decline   prevail   innumerable

1. The farmers ploughed the vegetable leaves back to _____ the soil.
2. We need to take corrective action to halt this country's _____.
3. A _____ is something that seems impossible but happens anyway.
4. Music is an _____ part of the school's curriculum.
5. He quoted some old Chinese sayings to _____ his points.
6. The Taj Mahal is a _____ building.
7. Barbaric customs still _____ in the mountainous area.
8. _____ books have been written on the subject.

III. **Translate the following English into Chinese or Chinese into English.**

1. architectural embellishments   6. 原始社会
2. Buddhist pagoda   7. 中亚文化
3. timber structures   8. 新篇章
4. masonry structures   9. 方形布局
5. stark cross axis   10. 洋房

Oral Practices

Conversation One

A: I understand you are studying architecture?

B: Yes, I really enjoy it.

A: What type of architecture is your favorite?

B: Well, I enjoy all types, but my favorite is architecture from the continent of Asia and particularly China.

A: Really? What is the main architectural philosophy in China?

B: Much of Chinese architecture is based on symmetry because it emphasizes balance. At the same time, Chinese architecture also uses curves and engravings on beams and lintels. The details are exquisite.

A: I think many continents have adopted some of the Chinese philosophies. I have seen pavilions supported by pillars in many other countries.

B: The wonderful thing about Chinese architecture is how it is based on ancient traditions, but has still been able to absorb new techniques and modern trends. It's been a pleasure to study and learn more about this fascinating style of architecture.

Conversation Two

A: I heard that you visited so many Chinese architectures this summer vacation?

B: Yes, I visited the Forbidden City, the Temple of Heaven, the Summer Palace and the Potala Palace.

A: That's wonderful. Which one do you like best?

B: Although the imperial palaces are very marvelous, I like the Potala Palace best.

A: Can you introduce it to me?

B: Of course. The Potala Palace is more than 117 meters (384 feet) in height and 360 meters (1,180 feet) in width, occupying a building space of 90 thousand square meters. It is composed of White Palace and Red Palace. The former is for secular use while the later is for religious.

A: What are the features as one of the most famous ancient Chinese architecture?

B: The design and construction of the palace are in accordance with the law of sunlight at high altitude. There are a variety of sculptures on columns and beams. Color murals on wall take up more than 2,500 square meters.

A: It sounds so great. I will visit it if there is any chance in the future.

B: Yes, you should go there and enjoy it.

Translation Skill—建筑英语翻译之语言结构特征

建筑英语文本同其他科技文本一样，严谨周密，概念准确，逻辑性强，行文简练，句式严整，少有变化。它的语言结构特点在翻译过程中如何处理，是进行英汉翻译时需要认真探讨的问题。现将从以下五个方面进行探讨。

1. 多名词化结构

为使行文简洁，科技英语中多用表示动作或状态的抽象名词或起名词作用的V-ing形式以及名词短语结构。

【例】Construction economy is a complicated subject involving raw materials, *fabrication*, *erection* and *maintenance*.

建筑经济是一个复杂的问题，其中包括原材料、制作、安装和维修。

【例】Concrete construction consists of several operations: *forming*, concrete *production*, *placement* and *curing*.

混凝土施工包括的几项操作有：支模板、混凝土生产、浇筑和养护。

2. 多长句和逻辑关联词

建筑英语中虽然大量使用名词化词语、名词短语结构以及悬垂结构来压缩句子长度，但是为了把事理充分说明，也常常使用一些含有许多短语和分句的长句，同时还常使用许多逻辑关联词，如hence, consequently, accordingly, then, however, but, finally, in short等，以使行文逻辑关系清楚、层次条理分明。

【例】The crane is the most common type of erection equipment, *but* when a structure is too high or extensive in area to be erected by a crane, it is necessary to place one or more derricks on the structure to handle the steel.

起重机是最普通的装配设备，但当结构太高或面积太大而无法安装时，有必要在结构上放置多个起重机来处理钢筋。

【例】Roofing is a larger part of building's periphery protecting structure, which suffers the highest outdoor temperature. *Therefore*, the importance of its energy-saving is particularly prominent. Roofing's warm-keeping is very important for improving indoor temperature environment and saving the overall energy consumption of high-rise building.

屋面是建筑外围保护结构所受室外温度最高的地方，面积也较大，因此，对高层建筑而言其节能的重要性尤为突出，屋面的保温隔热措施对改善高层建筑室内温度环境、节约整体能耗非常重要。

3. 多用一般现在时和完成时

现在时和完成时之所以在建筑英语中常见，是因为前者可以较好地表现文章内容的无时间性，说明文章中的科学定义、定理、公式不受时间限制，任何时候都成立；后者则多用来表述已经发现或获得的研究成果。

【例】The *planning* phase starts with a detailed study of construction plans and specifications.

施工计划开始于对建筑图纸和基本规范的详细研究。

【例】The Great Wall is the largest man-made monument ever to *have been built* and it is *disputed* that it is the only one visible from space.

长城是世界上最大的人造工程，据说也是唯一可以从外太空看到的地球景观。

4. 多用被动语态

建筑英语中被动语态的使用极为广泛，因此，英汉互译时，必须注意被动语态的翻译，不要过分拘泥于原文的被动结构，而要根据汉语的习惯灵活处理。

【例】Cost engineering is *defined* as that area of engineering practice where engineering judgment and experience *are utilized* in the application of scientific principles and techniques to the problem of cost estimation, cost control and profitability.

造价工程定义为：在运用科学原理和技术解决成本估算、成本控制和收益方面的问题时，需要运用工程判断和经验的工程实践领域。

【例】The ultimate strength of a material is *measured* by the stress at which failure（fracture）occurs.

材料的极限强度通过发生破坏（断裂）时的应力来测量。

5．多用非限定动词

建筑英语中，往往使用分词短语代替定语从句或状语从句；使用分词独立结构代替状语从句或并列分句；使用不定式短语代替各种从句；介词+动名词短语代替定语从句或状语从句。这样可缩短句子，又比较醒目。

【例】The Great Pyramid in Egypt, *standing* 481 feet（147 meters）high, is the most spectacular masonry construction.

埃及的大金字塔，481英尺①高（147米），是最壮观的砌体建筑。

【例】Materials *to be used* for structural purposes are chosen so as to behave elastically in the environmental conditions.

结构材料的选择应使其在外界条件中保持弹性。

6．多用后置定语

大量使用后置定语也是建筑文本的特点之一。常见的结构有以下四种：

（1）介词短语作后置定语。

【例】The soil *under a high building* often settles.

高层建筑物下面的土壤常常下沉。

（2）形容词及形容词短语作后置定语。

【例】In radiation, thermal energy is transformed into radiant energy, *similar in nature to light*.

热能在辐射时，转换成性质与光相似的辐射能。

（3）分词及分词短语作后置定语。

【例】Beam bridges are the simplest forms *supported by an abundant* at each end of the bridge deck.

梁桥是最简单的桥梁，桥面板的两端各有一个桥墩支撑。

（4）从句作后置定语。

【例】It has a high elastic modulus, *which results in small deformation under load*.

它有较高的弹性模量，其结果就是荷载作用下变形较小。

---

① 1英尺=0.304 8米。

Unit Three Building Design
建筑设计

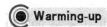

Task 1: Match the English expressions with their corresponding equivalents.

1. The Crooked House   A. 堪萨斯城公共图书馆
2. The Basket Building   B. 疯狂屋
3. Kansas City Public Library C. 集装箱城市
4. Cubic Houses   D. 篮子大楼
5. Crazy House   E. 颠倒屋
6. The UFO House   F. 飞碟屋
7. Container City   G. 扭曲屋
8. Upside-Down House   H. 立方体房子

Task 2: Match the pictures with the above famous architectures in Task 1.

1._____   2._____   3._____   4._____

5._____   6._____   7._____   8._____

Text A A Brief Introduction to the Building Design
建筑设计简介

The architect usually begins to work when the site and the type and cost of a building have been determined. Thus, designing is the process of **particularizing** and **harmonizing** the demands of environment, use, and economy. This process has a cultural as well as a **utilitarian** value, for in creating a plan for any social activity the architect **inevitably** influences the way in which that activity is **performed**.

Environmental Design

The natural environment is **at once** a **hindrance** and a help, and the architect seeks both to invite its aid and to **repel** its attacks. To make buildings **habitable** and comfortable, he must control the effects of heat, cold, light, air, **moisture**, dryness and foresee **destructive potentialities** such as fire, earthquake, flood and disease.

• **Orientation**

The arrangement of axes of buildings and their parts is a **device** for controlling the effects of sun, wind and rainfall[1]. The characteristics of the immediate environment (such as trees, land formations, bodies of water and other buildings) also influence orientation.

• Architectural Forms

The appropriate designing may control and **modify** the effects of natural forces. For example, **overhanging eaves** gives shade and protection from rain. Roofs are designed to shed snow and to drain or **preserve** water. Walls can **insulate** heat and control the amount of heat lost; Windows are the principal means of controlling natural light, ventilation and so on.

• Color

Color has a practical designing function as well to control the range of its reflection and its **absorption** of solar rays as an **expressive** quality[2].

• Materials and Techniques

The choice of materials **is conditioned by** their own ability to withstand the environment as well as by properties that make them useful to human beings[3]. Their advantages and disadvantages should be taken into consideration in design.

• Interior Control

Temperature, light and sound **are** all **subject to** control by the size and shape of interior spaces, the way in which the spaces are connected, the color and **texture** of materials employed for floors, walls, ceilings and **furnishings**[4]. Today, heating, insulation, air conditioning, lighting and **acoustical** methods have become basic parts of the architectural program.

Functional Design

While **environmental design** produces comfort for the sense (sight, feeling and hearing) and reflexes, functional design **is concerned with** convenience of movement and rest.

- Differentiation

Differentiate spaces for distinct functions.

- Circulation

Establish a proper route among different spaces and between the exterior and the interior, which must be simple, have evident goal and convenience of equipment.

- Facilitation

Analyze the body **measurements**, movements and muscular power of human beings for conditioning the measurements of ceilings, gates, windows and steps in order to make use of devices for **facilitating** the actions of the human body.

Economic Accounting Design

Major **expenses** in building are for land, materials and labor. When land coverage is limited it is usually necessary to design in height of the space[5]. When the choice of materials is influenced by cost, all **phases** of architectural design are affected. High labor costs encourage simplification in construction and the replacement of **craftsmanship** by **standardization**. Economic accounting designing involves not only proportioning of **expenditures** among land, materials and labor in order to produce the most effective solution to an architectural problem.

New Words

particularize [pəˈtɪkjələraɪz] vt. 特别指出，具体化
harmonize [ˈhɑːmənaɪz] vt. 调和，和谐
utilitarian [ˌjuːtɪlɪˈteərɪən] adj. 实用的，以实用为主的，功利的
inevitably [ɪnˈevɪtəbli] adv. 不可避免地，必然发生地
perform [pəˈfɔːm] vt. 执行，履行，做
hindrance [ˈhɪndrəns] n. 妨碍的人或物
repel [rɪˈpel] vt. 逐退，驱开
habitable [ˈhæbɪtəbl] adj. 适于居住的
moisture [ˈmɔɪstʃə(r)] n. 潮湿，湿气
destructive [dɪˈstrʌktɪv] adj. 毁灭性的
potentiality [pəˌtenʃɪˈælətɪ] n. 可能性，潜力

orientation [ˌɔːrɪənˈteɪʃn] n. 朝向，认识环境
device [dɪˈvaɪs] n. 装置，装置物
modify [ˈmɒdɪfaɪ] vt. 减轻，缓和
overhang [ˌəʊvəˈhæŋ] vt.&vi. (hung) 悬于……之上，悬垂
eaves [iːvz] n. 屋檐
preserve [prɪˈzɜːv] vt. 保留
insulate [ˈɪnsjuleɪt] vt. 隔热，绝缘
absorption [əbˈzɔːpʃn] n. 吸收
expressive [ɪkˈspresɪv] adj. 表现的
condition [kənˈdɪʃn] vt. 支配，决定，限制
subject [ˈsʌbdʒɪkt] adj. 服从的
texture [ˈtekstʃə(r)] n. 质地，结构
furnishing [ˈfɜːnɪʃɪŋ] n. 家具与陈设品（常用复数）

acoustical [əˈkuːstɪk] *adj.* 听觉的
concern [kənˈsɜːn] *vt.* 对……关系，对……有重要性
differentiation [ˌdɪfəˌrenʃɪˈeɪʃn] *n.* 区分，分别，辨别
differentiate [ˌdɪfəˈrenʃɪeɪt] *vt.* 区别，区分，辨别
facilitation [fəˌsɪlɪˈteɪʃn] *n.* 便利，便利设备
analyse [ˈænəlaɪz] *vt.* 分析，研究

measurements [ˈmeʒəmənts] *n. & pl.* 尺度，尺寸
facilitate [fəˈsɪlɪteɪt] *vt.* 使便利
expense [ɪkˈspens] *n.* 花费
phase [feɪz] *n.* 方面，片段，阶段
craftsmanship [ˈkrɑːftsmənʃɪp] *n.* 手艺，精巧的技艺
standardization [ˌstændədaɪˈzeɪʃn] *n.* 标准化
expenditure [ɪkˈspendɪtʃə(r)] *n.* 开销，经费

Phrases & Expressions

at once 同时
be conditioned by 受……限制，被……支配
be subject to 受控制于……

environmental design 环境设计
be concerned with 与……有关
economic accounting 经济核算

Notes on Text

[1] The arrangement of axes of buildings and their parts is a device for controlling the effects of sun, wind, and rainfall.

建筑物轴线的朝向以及内部空间布置是控制阳光、风雨影响的一种方法。

[2] Color has a practical designing function as well to control the range of its reflection and its absorption of solar rays as an expressive quality.

色彩有着控制吸收和反射日光程度的实用功能，同时，也有其美学上表现作用。

[3] The choice of materials is conditioned by their own ability to withstand the environment as well as by properties that make them useful to human beings.

材料选择时应同时考虑其本身防御自然条件的能力以及对人的利害关系。

[4] Temperature, light and sound are all subject to control by the size and shape of interior spaces, the way in which the spaces are connected, the color and texture of materials employed for floors, walls, ceilings and furnishings.

室内空间的大小、形状、连接方法，以及地面、墙面、天花板、家具等所用材料的颜色、质地均可以控制室内的温度、光线、音响等。

[5] Major expenses in building are for land, materials, and labor. When land coverage is limited it is usually necessary to design in height of the space.

建筑的费用主要用于土地、材料和人工。当土地有限时，建筑只能向高度发展。

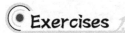 **Exercises**

Ⅰ. Answer the following questions according to the text.

1. What is the definition of design in our text?
2. What is the effect of the natural environment on design?
3. What is the scope of environmental design?
4. What is the scope of functional design?
5. What are the major expenses in building?

Ⅱ. Match the English words with their Chinese equivalents.

1. particularize   A. 毁灭性的
2. modify   B. 活动空间，环流
3. perform   C. 标准化
4. insulate   D. 使便利
5. moisture   E. 可能性，潜力
6. destructive   F. 特别指出，具体化
7. potentiality   G. 潮湿，湿气，水气
8. circulation   H. 减轻，缓和
9. standardization   I. 执行，履行
10. facilitate   J. 隔热，绝缘

Ⅲ. Read the following passage and fill in the blanks with the words given in the box.

| following | serviced | shift | design | effectively |
| low | maximum | local | curtain | granted |

Designing building, to _____ meet the conditions and realities of a Post-Carbon, Climate-Changed world will require a _____ in your current understanding of what constitutes good building design and sound building practice. Many of the practices that we now take for _____, like cladding our buildings in _____ wall building envelopes, in future, may no longer be economically feasible. To address these changes in building designing and construction, we have assembled the _____ building design principles for designing in a post-carbon, climate responsive building environment.

1. Use low carbon-input materials and systems.
2. _____ and plan buildings for low external energy inputs for ongoing building operations.

3. Design buildings for _____ daylight.

4. Design for future flexibility of use.

5. Design for durability and robustness（坚固性）.

6. Design for use of _____ materials and products.

7. Design and plan for _____ energy input constructability.

8. Design for use of building systems that can be _____ and maintained with local materials, parts and labor.

Ⅳ. **Translate the following English into Chinese or Chinese into English.**

1. utilitarian value
2. natural environment
3. destructive potentialities
4. immediate environment
5. natural light
6. 室内空间
7. 照明和音响效果
8. 主要花销
9. 功能设计
10. 经济核算

Text B Designers Redefining Modern Architecture
给现代建筑重新定义的建筑设计师

Robert Venturi, Frank Gehry, Tadao Ando and Zaha Hadid are four important building designers widely considered as the top **architects** at work today. They all have created important examples of modern architecture in very different ways[1]. You can see their **energizing** and **imaginative** contributions to modern designs in buildings around the world[2].

Robert Venturi

The writings and buildings of American architect Robert Venturi （Fig.3-1） have helped **redefine** the path of modern architecture since the 1960s. The Vanna Venturi house in **Philadelphia** was one of his first important projects.

In his writings and buildings, Robert Venturi **calls for** a kind of modern architecture that shows the influences of history while also including popular culture.

Fig.3-1 Robert Venturi

Examples of his buildings include the addition to the National Gallery in London, England. Venturi designed this large building which is located in **Trafalgar Square** and is one of the main buildings in the 19th Century, as a more modern verison. Venturi also designed buildings for many American colleges, including **Harvard University** and the University of

Pennsylvania.

Frank Gehry

Frank Gehry (Fig.3-2) believes that architecture is art. He has said that in some ways he has been more influenced by artists and **sculptors** than by architects. This may **be** why his buildings often look like energetic sculptures **made from** bold **geometric** forms.

Fig.3-2　Frank Gehry

His "Dancing House" is in **Prague**, the capital of the **Czech Republic**. This **playful** building, finished in 1996, looks like two dancers. Gehry's most famous building is the **Guggenheim Museum**, Bilbao in Spain which was completed in 1997. Most curvings of the building **are covered with titanium**. It looks like a dancing metal wave sitting on the edge of a river[3].

Tadao Ando

Tadao Ando (Fig.3-3) is a Japanese architect. In 1993, he won **Japan's Culture Design Prize** for the Rokko housing project. Ando is known for using unfinished reinforced concrete as a building material and making buildings with simple designs that **are** closely **connected to** the natural environment[4].

In 2002, his Modern Art Museum opened in Fort Worth, Texas. The building is so **striking** that you have to remind yourself to look at the art. The concrete and glass museum is built next to a body of water, so it almost seems to float.

Fig.3-3　Tadao Ando

Zaha Hadid

Born in Iraq, Zaha Hadid (Fig.3-4) completed her architectural studies in Britain, where she now lives. In 2004, Hadid became the first woman to win the important **Pritzker Prize** for architecture.

Hadid is known for designing and completing structures that seem almost impossible on paper. She says that architecture is clearly about **shelter**, but it also brings pleasure mostly. Her famous buildings include the car **park** and **terminus** in Strasbourg, France, the Rosenthal Center for **Contemporary** Arts in Cincinnati, Ohio, etc[5].

Fig.3-4　Zaha Hadid

New Words

architect [ˈɑːkɪtekt] n. 建筑师，设计师
energizing [ˈenədʒaɪzɪŋ] v. 充满活力的
imaginative [ɪˈmædʒɪnətɪv] adj. 富于想象力的
redefine [ˌriːdɪˈfaɪn] vt. 重新定义
sculptor [ˈskʌlptə(r)] n. 雕刻家
geometric [ˌdʒiːəˈmetrɪk] adj. 几何图形的
playful [ˈpleɪfl] adj. 有趣的

titanium [tɪˈteɪniəm] n. 钛
striking [ˈstraɪkɪŋ] adj. 显著的，突出的
shelter [ˈʃeltə(r)] n. 居所，避难所
park [pɑːk] n. 停车场
terminus [ˈtɜːmɪnəs] n. 终点，终点站
contemporary [kənˈtemprəri] adj. 当代的，现代的

Phrases & Expressions

call for 呼吁
be made from 由……所做成
be covered with 覆盖着

be connected to 与……有联系，与……有关联

Proper Names

Philadelphia 费城
Trafalgar Square 特拉法尔广场
Harvard University 哈佛大学
Prague 布拉格

Czech Republic 捷克共和国
Guggenheim Museum 古根海姆博物馆
Japan's Culture Design Prize 日本文化创意奖
Pritzker Prize 普利兹克建筑奖

Notes on Text

[1] They have all created important examples of modern architecture in very different ways.
他们都以不同的方式为现代建筑创造了重要的范例。

[2] You can see their energizing and imaginative contributions to modern designs in buildings around the world.
你可以在全世界的现代建筑设计中看到他们那充满活力和想象力的建筑。

[3] Most curvings of the building are covered with titanium. It looks like a dancing metal wave sitting on the edge of a river.
这座建筑的绝大部分弯曲处都覆盖着钛金属，它看起来就像是河畔边跳动的金属波浪。

[4] Ando is known for using unfinished reinforced concrete as a building material and making

buildings with simple designs that are closely connected to the natural environment.

安藤因把未成型的钢筋混凝土作为建筑材料而闻名，而且，他还因以尽可能地与自然环境相融合的简单设计来建造住房而闻名。

[5] Her famous buildings include the car park and terminus in Strasbourg, France, the Rosenthal Center for Contemporary Arts in Cincinnati, Ohio, etc.

哈迪德的建筑工程和她的建筑设计包括法国斯特拉斯堡停车场和终点站，俄亥俄州辛辛那提的罗森塔尔当代艺术中心等。

Exercises

Ⅰ. **Choose the best answers according to the text.**

1. Who has helped redefine the path of modern architecture since the 1960s?
   A. Robert Venturi.   B. Frank Gehry.   C. Tadao Ando.   D. Zaha Hadid.

2. The following buildings are designed by Robert Venturi except _____.
   A. National Gallery in London   B. University of Pennsylvania
   C. Harvard University   D. University of Cambridge

3. _____ is Frank Gehry's most famous building in Spain.
   A. Dancing House.   B. Guggenheim Museum.
   C. Chiat/Day Building.   D. New World Center.

4. What is TadaoAndo famous for?
   A. He is famous for making buildings with simple designs and unfinished reinforced concrete.
   B. He is famous for making buildings with complex designs and steel.
   C. He is famous for making buildings with simple designs and glass.
   D. He is famous for making buildings with complex designs and timber.

5. Who was the first woman to win the important Pritzker Prize for architecture?
   A. Frei Otto.   B. B. V. Doshi.   C. Zaha Hadid.   D. Wang Shu.

Ⅱ. **Fill in the blanks with the following words. Change the forms if necessary.**

| architect | imaginative | redefine | version |
| contemporary | playful | striking | terminus |

1. There is a _____ difference between Jane and Mary.
2. The new building was built from the design of a famous _____.
3. The passengers were transferred to a ferry at the bus _____.

4. With them she is so _____ and eloquent.

5. The mind of the storyteller has great _____ powers.

6. His _____ what happened is incorrect.

7. _____ cars are more streamlined than older ones.

8. Healthcare City will _____ the region's medical capabilities.

Ⅲ. **Translate the following English into Chinese or Chinese into English.**

1. top architects
2. modern architecture
3. Dancing House
4. Guggenheim Museum
5. Pritzker Prize in Architecture
6. 具有活力的雕刻品
7. 现代模式
8. 几何图形
9. 自然环境
10. 未成型的钢筋混凝土

Oral Practices

Conversation One

A: Dwellers have different demands in different times. Designers must take them into account.

B: Yes, that's true. In the seventies last century, a majority of families preferred single-family homes. The buildings in which they lived were only 3-5 stories high. But since the nineties, many multi-story multiple apartment buildings have been rising all over China.

A: What's the reason for it?

B: It is because the city land is too expensive to be used for small housing units. The same plot of ground that would hold only a few single-family units can house many families, which is more considerate indeed.

A: Many dwellers think it is inconvenient to live in houses of more stories, where life would be dull. How do you handle it?

B: Quite a few apartment houses offer city dwellers large balconies, where they can grow plants or eat outdoors. What is more, there are several elevators working together. Apartment houses also provide garage space, a park and a playground.

A: Now many families prefer to live as far out as possible from the center of a metropolitan area. Is it because of the air pollution?

B: Yes, and another aspect is that they want to get away from noise, crowding and confusion. As time goes on, more and more citizens drive their cars to work. The accessibility of public transportation has ceased to be a decisive factor in housing.

Conversation Two

A: How are you getting on with your first design?

B: Very badly. I find it is difficult.

A: Why?

B: As you know, the project I undertake is so complex. I'm always at a loss and find it impossible to do.

A: Have you had a talk with your client?

B: Yes, but my client couldn't express his requirements very well.

A: Well, maybe this is where the question lies.

B: What do you mean?

A: Whenever I try to start with my design, I will know exactly what my clients really need.

B: Do you mean I should have a further talk with my client?

A: Yes, of course. When chatting with your client, you're expected to understand your client as much as possible. For example, the interest, character, temper of your client. So you can understand what he really needs.

B: That's a good idea. Thank you for your advice.

Translation Skill—建筑英语翻译之词义的选择及引申

1. 词义的选择

就词义的选择来说，建筑英语词义选择面临以下难点：一是词汇趋于专业化；二是普通词汇的专业含义。词汇的专业化表现在建筑英语中有大量的专业性术语；普通词汇的专业含义。

（1）建筑英语专业词汇。

precast：预浇筑　　　mortar：砂浆　　　asphalt：沥青

（2）普通词汇的专业含义。

concrete：普通释义为"具体的"；建筑英语中译为"混凝土"。

plate：普通释义为"盘子"；建筑英语中译为"金属板"。

mold：普通释义为"塑造"；建筑英语中译为"浇筑"。

foundation：普通释义为"基础"；建筑英语中译为"地基"。

（3）根据词类选择词义。

【例】Concrete can be *pumped* into place by special concrete *pumps*.（pump 动词）

混凝土可以通过特殊混凝土泵浇筑于现场。

例句中的第一个"pump"是动词，译为"用泵浇筑"；第二个"pump"是名词，译为"泵"。

【例】Using *prestress* to eliminate cracking means that the entire cross section is available to resist bending.

利用预应力来避免裂缝的出现意味着整个截面可以抗弯。（"prestress"是名词，译为"预应力"）

【例】When a curved tendon is used to *prestress* a beam, additional normal force develops between the tendon and the concrete because of the curvature of the tendon axis.

当采用曲线钢筋束来对梁施加预应力时，由于钢筋束轴线的弯曲影响，在钢筋和混凝土之间会产生附加径向压力。（"prestress"是动词，译为"施加预应力"）

2. 词义的引申

专业英语英译汉时，有时会遇到某些词在词典中找不到适当的词义，如果生搬硬套词典中的释义，译文则不能确切地表达原文的意思，甚至造成误译。这时就应结合上下文和逻辑关系，根据汉语的表达习惯，引申词义。词义的引申主要包括三个方面的内容。

（1）专业化引申。

在建筑英语语境中赋予了词汇不同于日常语境的专业化语义。因此，翻译时应基于其基本意义，根据所涉及专业引申出其专业化语义。

【例】This kind of wood *works* easily.

这种木料易于加工。（work不译为"工作"）

【例】Concrete construction consists of several operations: *forming*, concrete production, *placement*, and *curing*.

混凝土施工包含以下几个操作：模板制作，混凝土生产、浇筑及养护。（forming，placement，curing不译为"形成""安置""食品加工法"）

（2）具体化或形象化引申。

在建筑英语中，有些字面语义颇为笼统或抽象的词语，若按字面译出，要么不符合汉语表达习惯，要么难以准确地传达原文所表达的意思。在这种情况下，就应该根据特定的语境，用比较具体或形象化的汉语词语对英文词语所表达的词义加以引申。

【例】Steel and cast iron also differ in *carbon*.

钢和铸铁的含碳量也不相同。（carbon不译作"碳"）

【例】The choice of material in construction of bridges is basically between steel and concrete, while main trouble with concrete is that its tensile *strength* is very small.

桥梁建筑材料基本上仍在钢材和混凝土之间选择，而混凝土的主要缺点是抗拉强度低。

（3）概括化或抽象化引申。

建筑英语文章中有些词语的字面意思比较具体或形象，但若直译成汉语，有时则显得牵强，不符合汉语的表达习惯，使人感到费解。在这种情况下，就应用含义较为概括或抽象的词语对英文词语所表达的词义加以引申。

【例】Alloys belong to a *half-way house* between mixtures and compounds. 合金是介于混合物和化合物之间的一种中间结构。（half-way house不译作"两地间中途歇脚的客栈"）

【例】Industrialization and environmental degradation seem to go *hand in hand*.

工业化发展似乎伴随着环境的退化。（hand in hand不译作"携手"）

Unit Four Building Materials
建筑材料

Warming-up

Task 1: Match the Chinese expressions with their corresponding English equivalents.

( ) 1. 有机材料　　　　　　　　A. materials for wall
( ) 2. 无机材料　　　　　　　　B. erection materials
( ) 3. 复合材料　　　　　　　　C. building function materials
( ) 4. 建筑结构材料　　　　　　D. inorganic materials
( ) 5. 墙体材料　　　　　　　　E. composite materials
( ) 6. 建筑功能材料　　　　　　F. welding materials
( ) 7. 安装材料　　　　　　　　G. fireproofing materials
( ) 8. 装饰材料　　　　　　　　H. organic materials
( ) 9. 防火材料　　　　　　　　I. decorative materials
( ) 10. 焊接材料　　　　　　　　J. building structural materials

Task 2: Study the following symbols of building materials and fill in the blanks.

Number	Symbols of Building Materials	Chinese Meanings	Number	Symbols of Building Materials	Chinese Meanings
1	natural soil		4	fire-resistant brick	
2	rammed earth		5	hollow brick	
3	sand, ash and earth		6	concrete	

Number	Symbols of Building Materials	Chinese Meanings	Number	Symbols of Building Materials	Chinese Meanings
7	sand gravel		12	reinforced concrete	
8	natural stone		13	multirole materials	
9	rubble		14	glass	
10	common brick		15	plastic	
11	wood		16	metal	

Text A A Brief Introduction to the Building Materials
建筑材料简介

Building material is any material which is used for construction purposes. Many naturally occurring **substances**, such as clay, rocks, sand, and wood, even **twigs** and leaves, have been used to construct buildings. Apart from naturally occurring materials, many man-made products are in use. In history there are trends in building materials from being: natural to becoming more man-made and **composite**; **biodegradable** to **imperishable**; **indigenous** (local) to being transported globally; repairable to **disposable**; and chosen for increased levels of fire-safety[1]. These trends tend to increase the initial and long term economic, **ecological**, energy and social costs of building materials.

Masonry

Masonry consists of natural materials, such as stone (Fig.4-1), or manufactured products, such as brick and **concrete** blocks. Masonry has been used since ancient times: mud bricks were used in the city of Babylon for **secular** building, and stone was used for the great temples of the Nile Valley. The Great Pyramid in Egypt is the most **spectacular** masonry construction.

Fig.4-1 Stone

Timber

Timber is one of the earliest construction materials and one of the few natural materials with good **tensile** properties. Hundreds of different **species** of wood are found throughout the world and each species exhibits different physical characteristics. Only a few species are used structurally as framing members in building construction.

Steel

Steel (Fig.4-2) is an outstanding structural material. It can be formed by rolling into various structural shapes such as I-beams, **plates**, and **sheets**; it also can be cast into complex shapes[2]. The addition of **alloying** elements results in higher-strength steels. These steels are used where the size of a structural member becomes **critical**, as in the case of columns in a skyscraper.

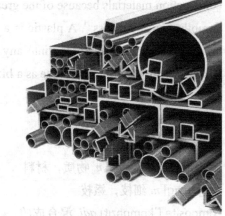

Fig.4-2 Steel

Aluminum

Aluminum (Fig.4-3) is especially useful as a building material when **lightweight**, strength, and **corrosion resistance** are all important factors[3]. Because pure aluminum is extremely soft and ductile alloying elements, such as **magnesium**, **silicon**, zinc, and copper, must be added to it to impart the strength required for structural use. Apart from its use for framing members in buildings and **prefabricated** housing, aluminum also finds extensive use for window frames and for the skin of the building in **curtain-wall construction**.

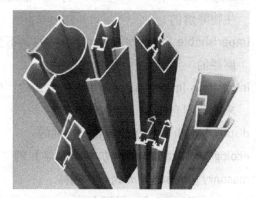

Fig.4-3 Aluminum

Concrete

Concrete is a mixture of water, sand, **gravel** and Portland cement. When concrete has been placed after mixing, it hardens into a dense and rocklike mass of great strength. It is basically a **compressive** material but has **negligible** tensile strength[4].

Reinforced Concrete

Concrete is strong in **compression**, but weak in **tension**. Steel has high tensile. When **reinforced** with steel bars, concrete will be strong both in compression and in tension. Such material is known as reinforced concrete. In construction, reinforced concrete can be **molded** into innumerable shapes, such as beams, columns, **slabs** and arches, and is therefore easily adapted to a particular form of building.

Plastics

Plastics (Fig.4-4) are rapidly becoming important construction materials because of the great variety, strength, durability and lightness[5]. A plastic is a **synthetic** material or **resin** which can be molded into any desired shape and which uses an **organic substance** as a **binder**.

Fig.4-4　Plastics

New Words

substance [ˈsʌbstəns] *n.* 物质，材料
twig [twɪɡ] *n.* 细枝，嫩枝
composite [ˈkɒmpəzɪt] *adj.* 混合成的
biodegradable [ˌbaɪəʊdɪˈɡreɪdəbl] *adj.* 能进行生物降解的
imperishable [ɪmˈperɪʃəbl] *adj.* 不朽的，不会腐烂的
indigenous [ɪnˈdɪdʒənəs] *adj.* 土生土长的，生来的，固有的
disposable [dɪˈspəʊzəbl] *adj.* 一次性的
ecological [ˌiːkəˈlɒdʒɪkl] *adj.* 生态（学）的
masonry [ˈmeɪsənri] *n.* 砌石，砌体
concrete [ˈkɒŋkriːt] *n.* 混凝土
secular [ˈsekjələ(r)] *adj.* 非宗教的，俗界的
spectacular [spekˈtækjələ(r)] *adj.* 壮观的

timber [ˈtɪmbə(r)] *n.* 木材，木料
tensile [ˈtensaɪl] *adj.* 拉力的，张力的
species [ˈspiːʃiːz] *n.* 物种，种类
plate [pleɪt] *n.* 金属板
sheet [ʃiːt] *n.* （金属材料）制成的薄板
alloy [ˈælɔɪ] *v.* 合铸，铸成合金
critical [ˈkrɪtɪkl] *adj.* 决定性的
aluminum [ˌæljəˈmɪnɪəm、ˌæləˈmɪnɪəm] *n.* 铝
lightweight [ˈlaɪtweɪt] *n.* 轻质，轻量级
magnesium [mæɡˈniːzɪəm] *n.* [化] 镁（金属元素）
silicon [ˈsɪlɪkən] *n.* <化>硅，硅元素
prefabricated [ˌpriːˈfæbrɪkeɪtɪd] *adj.* （建筑物）预制构件的
gravel [ˈɡrævl] *n.* 砂砾，碎石

compressive [kəmˈpresɪv] adj. 有压缩力的
negligible [ˈneglɪdʒəbl] adj. 可以忽略的，微不足道的
compression [kəmˈpreʃn] n. 压缩，紧缩
tension [ˈtenʃn] n. [物] 张力，拉力
reinforce [ˌriːɪnˈfɔːs] vt. 加固，使更结实，加强

mold [məʊld] vt. 浇铸，塑造
slab [slæb] n. 厚板，平板
synthetic [sɪnˈθetɪk] adj. 合成的，人造的
resin [ˈrezɪn] n. 树脂，合成树脂
binder [ˈbaɪndə(r)] n. 胶粘剂

Phrases & Expressions

corrosion resistance 耐腐蚀性
curtain-wall construction 幕墙结构

reinforced concrete 钢筋混凝土
organic substance 有机物质

Notes on Text

[1] In history there are trends in building materials from being: natural to becoming more man-made and composite; biodegradable to imperishable; indigenous（local）to being transported globally; repairable to disposable; and chosen for increased levels of fire-safety.

建筑材料的历史发展趋势从天然走向人工合成，从可降解到抗腐蚀，从本土使用到全球运输，从修复使用到一次性材料，并且选材的防火等级也不断提高。

[2] It can be formed by rolling into various structural shapes such as I-beams, plates, and sheets; it also can be cast into complex shapes.

它能通过碾压被做成不同的结构形式，比如I形梁、板和薄板；它还可以被浇铸成复杂的型式。

[3] Aluminum is especially useful as a building material when lightweight, strength, and corrosion resistance are all important factors.

当轻质、强度和抗腐蚀成为所有的重要因素时，铝就是一种特别有用的建筑材料。

[4] It is basically a compressive material but has negligible tensile strength.

混凝土基本上是一种抗压材料，但是抗拉强度是微不足道的。

[5] Plastics are rapidly becoming important construction materials because of the great variety, strength, durability and lightness.

塑料因为它的多样化、强度、耐久性和轻质性，很快就成为重要的建筑材料。

Exercises

I. Answer the following questions according to the text.

1. What is the developing trend of building materials in history?
2. What kinds of materials does masonry include?
3. What are the features of pure aluminum?
4. What is the difference between concrete and reinforced concrete?
5. Why do plastics become an important construction materials rapidly?

II. Match the English words with their Chinese equivalents.

1. substance   A. 砂砾，碎石
2. disposable   B. 合成的，人造的
3. concrete   C. [物] 张力，拉力
4. masonry   D. 合铸，铸成合金
5. timber   E. 一次性的
6. alloy   F. 砌石，砌体
7. gravel   G. 物质，材料
8. compression   H. 木材，木料
9. synthetic   I. 压缩，紧缩
10. tension   J. 混凝土

III. Read the following passage and fill in the blanks with the words given in the box.

| require | plastics | prime | stiffness | cost |
| materials | physical | manufactured | permanent | concrete |

Choosing different building _____ for the particular building application is a choice that designers must make. Natural materials, such as wood, stone and mud brick, are the oldest building materials, but new _____ materials, such as hard-burnt bricks and iron products are constantly being developed.

Building materials must have certain _____ properties（特性）to be structurally useful. Primarily, they must be able to carry a load, or weight, without any _____ change of shape. A second important property of a building material is its _____.

Fitness（适应性）for the intended purpose remains the _____ criterion（标准）, but _____ is often a determining factor. Thus, the common practice of replacing natural stone with brick and concrete is because of their lower cost.

While choosing materials, the concern about reducing the amount of energy required for heating and cooling buildings, and also a preference for materials and building manufacturing processes that _____ as little energy as possible are major considerations. Relatively cheap raw materials may be expensive to form. For example, the cost of the stone used in buildings today is mainly the cost of labor. In contrast, _____ can be cast into a mold without any cutting. _____ are in wide use because they can be easily formed into complex shapes at low cost and are easily mass produced.

Ⅳ. Translate the following English into Chinese or Chinese into English.

1. manufactured products
2. secular buildings
3. tensile properties
4. physical characteristics
5. structural material
6. 耐腐蚀性
7. 合金元素
8. 钢筋混凝土
9. 有机物质
10. 合成材料

Text B Little Ant Shadow Play Theater
小蚂蚁皮影剧场

Little Ant Shadow Play Theater is a public welfare project, located in Xiheyan Street, Beijing. The client is the Little Ant Shadow Play Troupe, consisting of patients with **growth hormone secretion deficiency**. The name "Little Ant" （Fig.4-5）comes from the fact that these **puppeteers** are only as tall as children, and look like children. The **shadow play troupe** is there to help improve the lives and self-esteems of these patients, but sadly they did not have a permanent theatre. Therefore, Dashilar, a traditional neighbourhood of craftsmen was where they chose to settle. It is not only an ideal home, but also a window for the **community** to pay more attention to these "little people."

Fig.4-5 Front door of the theatre

As a traditional handicraft, shadow play uses **intricate** and detailed carvings to produce charming figures used for the play. The design （Fig.4-6）**exploits** this characteristic of shadow play and applies the idea of detailed carvings into interior design. Golden **wheat straw board**, which has natural **texture** and a slight **fragrance** of wheat, was chosen to be the main material.

The **beams** combined with **vertical** carved boards extend to the ground on one sidewall, forming a big shelf. On the other hand, the upper half of the other sidewall was designed to be a full-length mirror, which **enlarges** the space two times its size. The stage was intended to **rotate** so that passage to the workshop in the back would be open when there is no performances[1]. Since the workshop is only two meters wide, a curved-profile table was designed, not only allowing the staff sit on both sides, but also allowing the master to guide their work in centre of desk.

Fig.4-6 Inside design

This public welfare project was supported through the support and donations of many social resources. The lighting was designed for free by the team who designed the 2008 Olympic Square. They also helped to acquire all the lamps for free. The entrance is composed of a large glass door, and that glass is the same as the ones used for the Apple Store **facades**. By having the world's highest **transparency**, the entrance facade would allow passers see the activities indoors clearly[2]. Almost all of material used in this project was a public donation, and that of course includes the architectural design (Fig.4-7).

Fig.4-7 Design concept

This design seeks to use **contemporary** design concept in order to create a new social stage for the disabled[3]. The Little Ant shadow play theatre received much praise from its occupants after its completion, where the public got to learn more about these beloved "little people." The architect also received warm responses when he **called on** everyone to support this welfare project. It also **testifies** that welfare projects can help **inspire positivity** to the society.

New Words

puppeteer [ˌpʌpɪˈtɪə(r)] n. 操纵木偶的人
community [kəˈmju:nəti] n. 社区；社会团体；共同体
intricate [ˈɪntrɪkət] adj. 错综复杂的；难理解的；曲折
exploit [ɪkˈsplɔɪt] vt. 开采；开拓；利用（……为自己谋利）；剥削
texture [ˈtekstʃə(r)] n. 质地；结构；本质；v. 使具有某种结构
fragrance [ˈfreɪɡrəns] n. 芳香；香气；香水(常用于广告语)
beam [bi:m] n. 梁，栋梁；束；光线 vt. 以梁支撑；播送；用……照射
vertical [ˈvɜ:tɪkl] adj. 垂直的，竖立的；n. [建] 竖杆；垂直线，垂直面

enlarge [ɪnˈlɑ:dʒ] vt.& vi. 扩大，放大；扩展，扩充
rotate [rəʊˈteɪt] vt. & vi. （使某物）旋转；使转动；使轮流，轮换
facade [fəˈsɑ:d] n. 建筑物的正面；外表
transparency [trænsˈpærənsi] n. 透明；透明度；透明的东西
contemporary [kənˈtemprəri] adj. 当代的，现代的 n. 同代人；当代人
testify [ˈtestɪfaɪ] vi. 证明；证实；作证 vt. 作证；证明，为……提供证明
inspire [ɪnˈspaɪə(r)] vt. 鼓舞；激励；赋予灵感 vi.吸，吸入
positivity [ˌpɒzəˈtɪvəti] n. 积极性

Phrases & Expressions

growth hormone secretion deficiency 生长激素分泌不足
shadow play troupe 皮影剧团
wheat straw board 麦秸板
call on 号召；要求；拜访（某人）

Notes on Text

[1] The stage was intended to rotate so that passage to the workshop in the back would be open when there is no performances.
舞台被设计为可旋转，这样便可以在非放映时间打开去往后面工作间的通道。

[2] By having the world's highest transparency, the entrance facade would allow passers see the activities indoors clearly.
世界上最高的透明度使内部的活动能最大限度地被过往的路人透视。

[3] This design seeks to use contemporary design concept in order to create a new social stage for the disabled.
这个设计力图用当下的设计理念，为一个特殊的残疾人群打造新的社会舞台。

Exercises

I. Choose the best answers according to the text.

1. What idea is applied in the design?
   A. Detailed carvings. B. Beautiful pictures.
   C. Traditional handicraft. D. Charming figures.

2. What main material was used in the design?
   A. Golden wheat straw board. B. Wooden structure.
   C. Natural marble. D. Steel and glasses.

3. A mirror was used to _____.
   A. enlarge the room B. enlarge the theatre
   C. enlarge the space D. enlarge the people

4. In this project, what were free mentioned in the paragraph?
   A. Lamps.   B. Glass.   C. Design.   D. All of the above.

5. What do the occupants think of the design?
   A. They don't like it at all. B. They think it is useful.
   C. They think it is ordinary. D. They praise it so much.

II. Fill in the blanks with the given words. Change the form if necessary.

| intricate | rotate | contemporary | texture |
| beam | facade | testify | vertical |

1. The earth _____ once every 24 hours.
2. The oriental rug has an _____ pattern.
3. _____ Indian movie has much to offer in its vitality and freshness.
4. The ceilings are supported by oak _____.
5. Each brick also varies slightly in tone, _____ and size.
6. For some buildings a _____ section is more informative than a plan.
7. Peter promised to _____ fully and truthfully.
8. The _____ of the building was a little weathered.

III. Translate the following English into Chinese or Chinese into English.

1. public welfare project　　　　6. 曲线轮廓的桌子
2. social resources　　　　　　　7. 通长的镜面

3. interior design
4. vertical carved boards
5. a traditional handicraft
8. 残疾人
9. 当代设计理念
10. 主材料

Oral Practices

Conversation One

A: Good morning!

B: Good morning!

A: As you see, the foundations are ready.

B: Yes, but what's your plan of supplied materials?

A: Firstly, we need a large number of bricks for laying walls, and other materials for the main structure should also be delivered. Secondly, installation and decorative materials should be sent to the worksite, such as water pipes, electric wires, marble and paints for decoration.

B: That's a good plan, but can all the materials be delivered on time?

A: No problem.

Conversation Two

A: We haven't seen each other for a long time, Miss Gao. Can I help you?

B: Long time no see! I am here to buy some building materials.

A: What materials do you want to buy?

B: Mainly common materials, such as cement, steel bar, timber and so on. Can you tell me their current market prices?

A: Oh, let me have a look. Cement is 2 dollars per kilogram, steel bar is 500 dollars per ton, and timber is 700 dollars per cubic meter.

B: Will you show me the price catalogue?

A: OK, here it is. The materials here are not expensive, but of the best quality. You can rest assured.

B: Honestly speaking, the prices are much higher than they were one year ago. I want to know the most favorable price for 20 tons of cement, 500 tons of steel bars and 500 cubic meters of timber?

A: Considering the quantity of these three building materials, we can offer you 10% discount.

B: Deal! Please deliver the materials on time.

A: We will deliver them to your site without any delay.

B: Thank you so much!

Translation Skill—建筑英语翻译之词义的增译和减译

由于英汉两种语言在语法结构、词类及修辞手段方面存在很大差异,在建筑英语的实际翻译过程中很难做到词汇和句法的完全对应。因此,为了准确传达原文信息,在翻译过程中往往需要对一些词汇进行增减,以达到翻译效果。

1. 增译

(1) 增加表示名词复数概念的词语。

翻译时,有时为了明确原文的含义,需要通过增译"们""一些""许多"等词语把英语中表示名词复数的概念译出。

【例】There is no enough steel to meet the world's construction needs for centuries to come.

并没有足够的钢材来满足全世界未来几个世纪的建筑需要。

【例】Carbon combines with oxygen to form carbon oxides.

碳同氧气化合形成多种氧化碳。

(2) 增加表示时态的词语。

汉语的动词没有表示时态的词形变化和相应的助动词,因此翻译时应增译相应的时间副词或助词,用来表示不同的时态。

【例】Contemporary building materials are now working for new important breakthroughs.

当代建筑材料正在酝酿着新的重大突破。

【例】This natural approach to remediating hazardous wastes in the soil, water and air is capturing the attention of landscape designers.

这种清除土壤、水以及空气中有害废品的自然方法正日益受到园林设计者的关注。

(3) 增加表示句子主语的词语。

当被动句中的谓语是表示"知道""了解""看见""认为""发现""考虑"等意思的动词时,通常可在其前增加"人们""我们""有人"等词语,译成汉语的主动句。

【例】This steel alloy is believed to be the best available here.

人们认为这种合金钢是这里能提供的最好的合金钢。

【例】It's believed that we shall make full use of the sun's energy to decrease the environmental pollution.

我们相信,总有一天我们将能够充分利用太阳能来减少环境污染。

(4) 增加原文中省略的词语。

英语句子的某些成分如果已在前面出现,有时则往往省略,但英译汉时,一般需要将其补出。

【例】The largest foundation sinks the most and the smallest the least.

最大的基础沉陷最多,而最小的基础沉陷最少。

【例】Starting with the largest sizes, boulders are larger than 10 cm, cobbles are from 5 cm to 10 cm.

从最大的粒径开始，漂石的粒径大于10 cm，卵石的粒径在5～10 cm。

（5）增加具体化、明确化的词语。

有些英语句子如果直译成汉语，意思表达不够具体和明确，因此汉译时须增译相关词语。

【例】From the 20th century, many building designers had already studied the possibility of zero-carbon construction.

从20世纪起，很多建筑设计师就已经开始研究建造零排放建筑。

【例】Low-carbon design is the first step toward the solution of air pollution.

低碳设计是解决空气污染问题的第一步。

2．减译

（1）省略冠词。

【例】*A* beam bridge is the simplest type of span.

梁桥是最简单的桥跨结构。

【例】*The* approach will avoid errors arising from assuming that the ultimate moments of resistance exist simultaneously at all the critical sections.

该方法将避免由于假定极限抵抗弯矩同时存在于所有临界截面而引起的误差。

（2）省略代词。

【例】If you know the internal forces, *you* can determine the proportion of members.

如果知道内力，就能确定构件尺寸。

【例】*It* is evident that a well lubricated bearing turns more easily than a dry one.

显然，润滑好的轴承比不润滑的轴承容易转动。

（3）省略动词。

【例】All kinds of excavators *perform* basically similar functions but appear in a variety of forms.

各种挖土机的作用基本相同，但形式不同。

【例】Concrete-block walls *have* very little lateral strength.

混凝土砌体墙的横向强度很低。

（4）省略介词。

【例】The difference in height measured in inches *between* the top *of* the concrete and the top *of* the slump cone is called the slump.

以英寸为单位测量的混凝土顶面与坍塌落度锥筒顶面的高差称为坍落度。

【例】The Old London Bridge was built *across* the Thames in 1209.

1209年，老伦敦桥建于泰晤士河上。

（5）省略连词。

【例】In downstream areas where the river passes *through* a broad gentle flood plain, civil engineers may be asked to build flood protection works.

在下游地区，河流经过的广阔平缓的洪泛平原，需要土木工程师修建防洪工程。

（6）省略意义上重复的词。

【例】The mechanical energy can be changed back into electrical energy by means of a *generator* or *dynamo*.

机械能可利用发电机再转变成电能。

Unit Five Building Structure
建筑结构

Warming-up

Task 1: Match the English expressions with their corresponding equivalents.

1. Arc de Triomphe   A. 白宫
2. Oriental Pearl TV Tower   B. 泰姬陵
3. Eiffel Tower   C. 迪拜帆船酒店
4. Beijing Bird's Nest   D. 比萨斜塔
5. White House   E. 东方明珠电视塔
6. Burj Al Arab   F. 凯旋门
7. Taj Mahal   G. 北京鸟巢
8. Leaning Tower of Pisa   H. 埃菲尔铁塔

Task 2: Match the pictures with the above famous architectures in Task 1.

1._____   2._____   3._____   4._____

5._____   6._____   7._____   8._____

Text A A Brief Introduction to the Building Structures
建筑结构简介

The part of the building that carries the weight and load is called the structure part. Parts, such as windows, that do not hold up the building are the non-structural parts. Building structures are classified into many forms according to different materials, such as reinforced **concrete structure**, **masonry structure** and steel structure.

Reinforced Concrete Structures (Fig.5-1)

In reinforced concrete structures, steel reinforcing bars **are embedded in** concrete structures where tensile stress may occur to make the good compressive strength of concrete structures fully put into action[1].

Reinforced concrete structures possess the features like large **dead mass**, high stiffness and good durability, etc. Because of

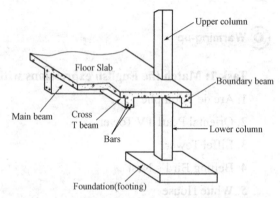

Fig.5-1 Typical reinforced concrete framing structure

the ways that buildings are made to hold up weight, which can be divided into different types, such as **framed** structure, shear wall structure, wall-framed structure and tube structure。

• **Framed structure** (Fig.5-2)

Bearing structural systems composed of elements that are longer than their **cross-sectional** dimensions are referred to as framed structure, such as beam and column[2].

• **Shear wall structure** (Fig.5-3)

Concrete continuous vertical walls may serve **architecturally** as partitions and it may also carry the gravity and lateral loading structurally. In this structure, shear walls **are** entirely **responsible for** the lateral load **resistance** of the buildings.

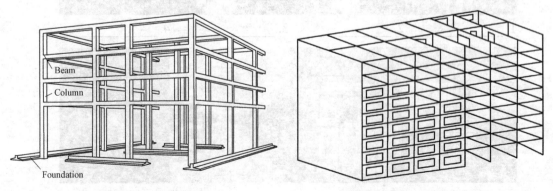

Fig.5-2 Framed structure   Fig.5-3 Shear wall structure

- **Wall-frame structure** (Fig.5-4)

When shear walls **are combined with** frames, the walls tend to **deflect** in a flexural **configuration**, and the frame tends to deflect in a shear mode. Both of them **are constrained** to adopt a common deflected shape by the **horizontal** rigidity of the **girder** and slabs[3]. The walls and frames interact horizontally, especially at the top, to produce a stiffer and stronger structure.

- **Tube structure** (Fig.5-5)

Only if all column **elements** can be connected to each other, can the maximum efficiency of the total structure of a tall building be **achieved** for both strength and **stiffness** to resist wind load[4]. The entire building acts as a **hollow** tube in projecting out of the ground, which is called tube structure.

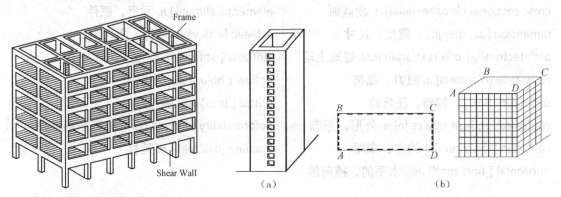

Fig.5-4 Wall-frame structure   Fig.5-5 Tube structure

Masonry structure (Fig.5-6)

The earliest use of masonry can **be traced back to** two thousand years ago in China, and masonry has a good quality of heat preservation and it's easy to construct, which is being used as a main material for civil buildings. But masonry material such as brick or block is a type of **brittle** material. Its **compressive capacity** is strong, however, its tensile capacity is as weak as its **deformability** or **ductility**[5].

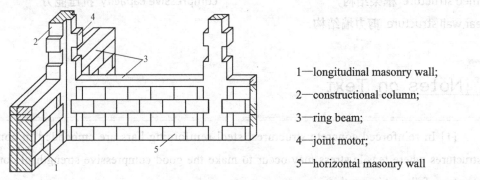

1—longitudinal masonry wall;
2—constructional column;
3—ring beam;
4—joint motor;
5—horizontal masonry wall

Fig.5-6 Masonry structure

Steel Structure

The steel structure refers to a broad one of building structures in which steel plays the leading role. Steel offers much better compression and tension than concrete and enables construction to be lighter. Steel structures use three-dimensional trusses, so they can have a larger span than reinforced concrete.

New Words

structure ['strʌktʃə(r)] n. 结构
frame [freɪm] n. 框架
cross-sectional ['krɒs'sekʃənəl] n. 横截面
dimension [daɪ'menʃn] n. 维度，尺寸
architecturally [ˌɑːkɪ'tektʃərəli] adv. 建筑上地
resistance [rɪ'zɪstəns] n. 阻力，抵抗
deflect [dɪ'flekt] v. 倾斜，使弯曲
configuration [kənˌfɪɡə'reɪʃn] n. 外形，形态
constrain [kən'streɪn] v. 约束，束缚
horizontal [ˌhɒrɪ'zɒntl] adj. 水平的，横向的

girder ['ɡɜːdə(r)] n. 主梁，纵梁
tube [tjuːb] n. 筒体
element ['elɪmənt] n. 元素，原件
achieve [ə'tʃiːv] vt. 取得，实现
stiffness ['stɪfnəs] n. 硬度
hollow ['hɒləʊ] adj. 中空的
brittle ['brɪtl] adj. 脆性的
deformability [dɪfɔːmə'bɪlɪti] n. 可变形性
ductility [dʌk'tɪləti] n. 延性

Phrases & Expressions

concrete structure 混凝土结构
masonry structure 砌体结构
be embedded in 嵌入
dead mass 自重
framed structure 框架结构
shear wall structure 剪力墙结构

wall-framed structure 框架剪力墙结构
be responsible for 对……负责
be combined with 与……联合
be traced back to 追溯到……
compressive capacity 抗压能力

Notes on Text

[1] In reinforced concrete structures, steel reinforcing bars are embedded in concrete structures where tensile stress may occur to make the good compressive strength of concrete structures fully put into action.

在钢筋混凝土结构中，利用钢筋被裹在混凝土结构中抵抗较大拉应力来使混凝土良好的抗压强度得以充分发挥。

[2] Bearing structural systems composed of elements that are longer than their cross-sectional dimensions are referred to as framed structure, such as beam and column.

由梁、柱构件通过刚性节点连接而成的超过横截面尺寸的承重结构称为框架结构。

[3] When shear walls are combined with frames, the walls tend to deflect in a flexural configuration, and the frame tends to deflect in a shear mode. Both of them are constrained to adopt a common deflected shape by the horizontal rigidity of the girder and slabs.

当剪力墙结构与框架结构结合，剪力墙就接近弯曲变形，而框架则趋于剪切变形，它们都是被大梁和板的水平刚度所约束而表现出来的共同的挠曲形态

[4] Only if all column elements can be connected to each other, can the maximum efficiency of the total structure of a tall building be achieved for both strength and stiffness to resist wind load.

只有将所有竖向构件相互连在一起，高层建筑所有结构类型中用于抵抗风荷载的强度和刚度才能达到最优值。

[5] Its compressive capacity is strong, however, its tensile capacity is as weak as its deformability or ductility.

它的抗压能力强，但拉伸能力弱，变形性或延性都较差。

Exercises

Ⅰ. Answer the following questions according to the text.

1. How many types can reinforced concrete structure be divided into? What are they?
2. When and where can the earliest use of masonry be traced back to?
3. Why is masonry being used as a main material for civil buildings?
4. What's the disadvantage of masonry material?
5. What makes the steel structures have a larger span than reinforced concrete?

Ⅱ. Match the English words with their Chinese equivalents.

1. structure   A. 主梁，纵梁
2. hollow   B. 框架
3. horizontal   C. 阻力，抵抗
4. cross-sectional   D. 建筑上地
5. dimension   E. 硬度
6. girder   F. 结构
7. stiffness   G. 维度，尺寸
8. architecturally   H. 横截面
9. resistance   I. 水平的，横向的
10. frame   J. 中空的

Ⅲ. Read the following passage and fill in the blanks with the words given in the box.

| cube | reflects | despite | frame | randomly |
| entirely | ETFE | as opposed to | enormous | plastic |

The Beijing Water Cube is a building for the Beijing 2008 Olympics, whose structure _____ the ancient. The building, which will be the National Aquatics Center, is indeed in the shape of a _____. It is made of a steel honeycomb-like _____ covered in a unique skin that is modeled after soap bubbles. Simply put, the effect is that the Water Cube looks like an _____ cube-shaped bundle of bubbles. The bubbles are made of a _____ called _____, which is also used to protect spaceships from cosmic radiation. One of the advantages of ETFE is that it traps solar energy in the winter and _____ it in the summer, helping to control the building's temperature. 3,500 bubbles had to be cut individually and sized in order to create the honeycomb-like structure. The bubbles are not identical or symmetrical, but seem to be organized _____, with different shapes and sizes nestled together. _____ its random appearance, however, the soap-bubble structure used in the design has a geometry that's perfect for a high-tech building. Soap bubbles actually always cling together in regular patterns, and the fragile-looking skin of the building's bubbles—the plastic covering is only 1/5 of a millimeter thick!—is _____ safe.

Ⅳ. Translate the following English into Chinese or Chinese into English.

1. framed structure
2. concrete structure
3. masonry structure
4. horizontal masonry wall
5. main beam

6. 筒体结构
7. 剪力墙结构
8. 框架-剪力墙结构
9. 抗压能力
10. 自重

Text B   Cattle Back Mountain Volunteers' House
牛背山志愿者之家

The Cattle back mountain volunteers' house (Fig.5-7) is a social **welfare** project **dedicating** to building a house for young volunteers in the mountain of Pumaidi Village, Luding County, Sichuan Province. Known as the best viewing platform for clouds, Niubeishan, meets a large number of travelers and explorers every year, yet, due to the short of resource and development, public **infrastructure** and social support is still rather **lagged behind**. Pumaidi Village, the closest village from Niubeishan that **inhabits** villagers, presents itself as a typical village in Southwest China, with **pitched** roof, green **tile**, and **harmonious** living environment. Like most of the villages in suburban area in China, the **majority** of the population **is composed of** the elderly and children, while the young and strong are out serving in the big cities trying to **make a living** for home. And thus, most of the houses in the village have been standing for long and **maintenance** is in **desperate** need. Considering all these, Dayan, an experienced volunteer, decided to build a social project **base** here, not only to provide help for those travelers in need, but also the elderly and children in the village. The house will serve as a **Youth Hostel** at the same time so as to keep the financial balance of the social project.

Fig.5-7　**Building Panorama**

Upon renovation, it is an old traditional folk house, with wood pitched roof and broken tiles, facing with a platform (locally named as "Bazi") in front, that is fragmented and shaded as couple dark rooms by couple thick walls. The loft on the rooftop is old and **shabby**, with no kitchen or bathroom inside. On the south of the platform "Bazi", there is a squared brick house made by farmers themselves, that's neither **coordinate to** the surrounding nor **resistant to** earthquake.

Our design (Fig.5-8) **strategy** is while improving the basic programs and functions, to make the architecture more open and public, capable of serving more people, and **blend** in the new building equipped with architectural and structural innovation with the surrounding environment and it straditional culture.

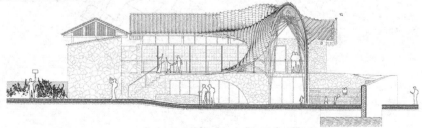

Fig.5-8　**Design concept**

We kept and strengthened the internal wood structure, got rid of the thick walls facing with the Bazi and its partitions to open up the first floor (Fig.5-9) and serve as an important public space where people can read, meet and grab a drink. The steel net framed glass wall can be used to store firewood, and when it open completely, it **merges** the interior and exterior **into** one.

Fig.5-9　Outside view of the first floor

The old sty on the north of the Bazi is removed, while the wood structure and the roof are kept to **transform into** a kitchen and a bathroom, with the only proper **flush** toilet in the whole village **installed** (Fig.5-10).

We also removed the brick house on the south of the Bazi, and replaced it with a pavilion with wood structure and tiles on top

Fig.5-10　Pitched roof

that **shelters** people from wind and rain[1]. With all the above, we **maximize** the use of the Bazi, and also form a unique viewing platform.

In the project, we used the most basic architecture materials and the most common building technique (stone walls, pitched roofs and green tiles that consists with the original language), and we try our utmost to make full use of the local human resource. Modern **digital** logic and design strategies are also **implemented** in the process (Fig.5-11) [2].

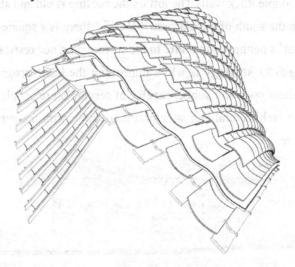

Fig.5-11　Local building technique

Looking at the main **facade**, from left to right, you will be able to see an **organic** roof shape merging with the mountain and clouds at its back, and the perfect **transition** from traditional art to modern art, or even the exploration of the future.

As for the internal space, it's a brand new expression of wood structure in digital times. Another part worth mentioning, the wood material used in the project is produced locally in Sichuan, a new type of bamboo-based **fiber composite**（Fig.5-12）with high resistance to strength, **moisture** and fire that's also **recyclable** and environmentally friendly.Being the first time for the producers to implement the material to irregular shaped architecture like this one, the producers took high participation in the process involving sampling on site, custom pre-made in the factory and manual adjustment on site.

Fig.5-12　Bamboo-based fiber composite

In talking about the special design of this part, principal architect Daode Li stated: The weaving roof and the mountain, along with the clouds at its back are connected in a way **aesthetically**, but yet, we hope this connection can extend to a more **spiritual** level. When traveler, volunteers, or even villagers approach from afar, looking at this unique and familiar building **shimmering** with warmth, they are then seized by the feeling of belongings, like a long **drifting** ship in the ocean which spots a light tower that gives them immediate courage to move forward.

New Words

welfare ['welfeə(r)] n. 福利；幸福；繁荣；安宁
dedicate ['dedɪkeɪt] vt. 奉献，献身
infrastructure ['ɪnfrəstrʌktʃə(r)] n. 基础设施；基础建设
inhabit [ɪn'hæbɪt] vt. 居住；在……出现；填满；vi. 居住
pitched [pɪtʃt] adj.（屋顶）有坡度的；v. 投（pitch的过去式和过去分词）；用沥青涂；排列
tile [taɪl] n. 瓦片，瓷砖；空心砖；麻将牌；vt. 用瓦片、瓷砖等覆盖

harmonious [hɑː'məʊniəs] adj. 和谐的，融洽的；悦耳的
majority [mə'dʒɒrəti] n. 多数；（获胜的）票数；成年；法定年龄
maintenance ['meɪntənəns] n. 维持，保持；保养，保管；维护；维修
desperate ['despərət] adj. 绝望的；铤而走险的；极度渴望的
base [beɪs] n. 基础；基地；根据；vt. 基于；把……放在或设在（基地）上
youth hostel ['juːθ hɒstl] 青年招待所，青年（学生）宿舍

shabby [ˈʃæbi] adj. 破旧的；卑鄙的；悭吝的
strategy [ˈstrætədʒi] n. 策略，战略；战略学
blend [blend] vt. 混合；（使）调和 vi. 掺杂；结合；n. 混合；混合物
flush [flʌʃ] vi. 奔流；冲刷 vt.（以水）冲刷，冲洗；n. 奔流，涌出
install [ɪnˈstɔːl] vt. 安装；安顿，安置
shelter [ˈʃeltə(r)] n. 遮蔽；居所 vt. 掩蔽；庇护；vi. 躲避；避难
maximize [ˈmæksɪmaɪz] vt. 最大化，使（某事物）增至最大限度
digital [ˈdɪdʒɪtl] adj. 数字的；数据的
implement [ˈɪmplɪment] vt. 实施，执行；使生效，实现
facade [fəˈsɑːd] n. 建筑物的正面；外表；虚伪，假象

organic [ɔːˈɡænɪk] adj. 有机（体）的；有组织的，系统的
transition [trænˈzɪʃn] n. 过渡，转变，变迁
fiber [ˈfaɪbə(r)] n. 光纤；（织物的）质地；纤维，纤维物质
composite [ˈkɒmpəzɪt] adj. [建] 综合式的；混合成的；n. 合成物，复合材料
moisture [ˈmɔɪstʃə(r)] n. 水分；湿气；潮湿；降雨量
recyclable [ˌriːˈsaɪkləbl] adj. 可循环再用的
aesthetically [iːsˈθetɪkli] adv. 审美地，美学观点上地
spiritual [ˈspɪrɪtʃuəl] adj. 精神的；心灵的；高尚的
shimmer [ˈʃɪmə(r)] vi. 闪烁发微光 n. 微光；闪光
drift [drɪft] v. 流动，漂流；浮现 n. 漂移

Phrases & Expressions

lag behind 落后，退步
be composed of 由……组成
make a living 赚钱过活，谋生
coordinate to 与……协调

resistant to 对……有抵抗力
merge into 汇合，（使）并入；归并
transform into 把……转变成……

Notes on Text

[1] We also removed the brick house on the south of the Bazi, and replaced it with a pavilion with wood structure and tiles on top that shelters people from wind and rain.
我们拆除了坝子南侧后建的砖房，还原了坝子原本的空间，并加建了一个木结构的构筑物，顶部覆瓦，可遮风避雨。

[2] Modern digital logic and design strategies are also implemented in the process.
在加建的构筑部分我们采用了数字化的设计方法与生成逻辑。

Exercises

I. **Choose the best answers according to the passage.**

1. From the first paragraph, we know the purpose of this project is to_____.
   A. help travelers
   B. build a social project base
   C. help the local people
   D. All of the above

2. From Paragraph Two, we can know this project is about_____.
   A. building a new house
   B. renovating an old house
   C. building a youth hostel
   D. renovating an old hostel

3. The strategy of this project is to_____.
   A. improve the basic functions
   B. connect with the surrounding environment
   C. make the architecture more public
   D. all of the above

4. What was the first step of the project?
   A. Kept and strengthened the internal wood structure.
   B. Got rid of the thick walls.
   C. Open up the first floor.
   D. Make a steel net framed glass wall.

5. What was built to replace the brick house?
   A. A new brick house.
   B. A new brick pavilion.
   C. A new wood house.
   D. A new wood pavilion.

6. What's the purpose of using local building materials?
   A. To save money.
   B. To combine with local people.
   C. To get money.
   D. To combine with local nature.

7. What was the newest point of the project?
   A. The new method.
   B. The new material.
   C. The new design.
   D. The new people.

8. What did the architect think of the project?
   A. It's an innovation of the time.
   B. It's a connection between nature and people.
   C. It's a creation of the designer.
   D. It's an imagination of the people.

II. **Fill in the blanks with the following words. Change the forms if necessary.**

| material | interior | recyclable | pitched |
| infrastructure | transform | pre-made | structure |

1. In its untreated state the carbon fibre_____ is rather like cloth.
2. _____ decoration by careful coordination seems to have had its day.

3. Motion Pants is manufactured by recycled fiber and is completely _____.

4. A _____ roof is one that slopes as opposed to one that is flat.

5. Vast sums are needed to maintain the _____.

6. The sofa can _____ for use as a bed.

7. He focuses on cement _____ parts.

8. The theatre is a futuristic steel and glass _____.

Ⅲ. **Translate the following English into Chinese or Chinese into English.**

1. public infrastructure
2. resistant to earthquake
3. sampling on site
4. an old traditional folk house
5. viewing platform

6. 冲水厕所
7. 建筑结构创新
8. 主立面
9. 异形建筑
10. 预先定制

Oral Practices

Conversation One

A: Good morning, Mr. Green.

B: Good morning, Jack. I'd like to know what kind of structure will be used in the new shopping center.

A: Well, we are still discussing. Someone suggests to use traditional styles like framed structure and tube structure, while others tend to use trusses, shell or membrane structures.

B: I can see that you are all creative and passionate about the project.

A: Yes, we wish the design will turn out to be architecture of fashion as well as stability, strength, stiffness and economy.

B: Good. Carry on and don't forget to tell me your final proposal next week. You know I have to show the design to the client after you hand in your work.

A: I see, Mr. Green. We will continue to finish the design and give you the final proposal and the sketch of it on time.

B: Well, goodbye.

A: Goodbye, Mr. Green.

Conversation Two

A: The structure is a very important part of the works. Now, let's go and have a close look at the fabrication of the steel structure step by step.

B: That's good. Let's have a look at the erection of the steel structure. The principle sequence of erection works is from low level to high level, from column to beam, from the main beams to secondary beams and from centers to outer sides.

A: All columns and beams are connected by high strength bolts. It's not allowed to have a single bolt fixed by over force. How can you manage it?

B: The sizes and position of the holes of the connection molds are controlled below 25% the tolerance of the specification. That allows a limitation of erection error. And we adopt that high accuracy connection molds during the fabrication of steel structure.

A: Are all the molds made in the same factory?

B: Yes, they are.

A: I have another question. Can you make the shop drawings if we provide you the basic design drawing?

B: Yes, we can. We have a design office equipped for CAD. We have specialists to make drawings like this. They are qualified engineers.

A: That's fine. Thanks a lot.

B: With pleasure.

Translation Skill—建筑英语翻译之定语从句的翻译

定语从句在建筑英语中的应用很广。由于英语中定语从句有长有短,对先行词的限制有强有弱,翻译时就必须根据每个句子的特点,结合上下文灵活处理。一般来说,定语从句在逻辑意义上往往与所限定的词有着"目的""结果""原因""让步"等联系。同时,在翻译定语从句时,一定要考虑到汉语的表达习惯。建筑英语中常见的定语从句翻译有以下几种方法。

1. 译成前置定语

限定性定语从句往往译成前置定语结构,即译成"……的"。但有些非限定性定语从句有时也可以做前置处理,尤其是当从句本身较短,或与被修饰词关系较为密切,或因拆译造成译文结构松散时。

【例】In the design of concrete-structures, an engineer can specify the type of material *that he will use*.

在混凝土结构设计中,工程师可以指定他将要使用的材料品种。

【例】Stair is the vertical transport facilities *which contact the two interfacing floors in the building*.

楼梯是联系建筑物上下层的垂直交通设施。

【例】The structure means the system *that can bear and transfer loads*, including rods, plates, shells and their combinations, such as bridges, roof trusses and so on.

工程结构是指能够承受和传递外部荷载的系统,包括杆、板、壳以及它们的组合体,如桥梁、屋架等。

2. 译成并列句

将定语从句翻译在所修饰的先行词后面，翻译为并列分句。

【例】Their main interest is in large dams, *where they may reduce the heat given out by the cement during hardening*.

它们主要用于大型水坝，在大坝中它们能减少水泥硬化时释放出的热量。

【例】The tendons are frequently passed through continuous channels formed by metal or plastic ducts, *which are positioned securely in the forms before the concrete is cast*.

预应力钢筋束穿入金属管或塑料管制成的连续孔道，而这些金属或塑料管在浇筑混凝土之前被固定在模板之中。

【例】Different structural systems have evolved from residential and office buildings, *which reflect their different functional requirement*.

住宅及办公楼建筑物常采用不同的结构体系，这反映出该类建筑不同的功能要求。

3. 译成谓语

当关系代词在定语从句中充当主语且句子的重点是在从句上时，可以省去关系代词，而将定语从句的其余部分译为谓语结构，以先行词充当它的主语，从而使先行词与定语从句合译成一句。

【例】A code is a set of specifications and standards *that control important technical specifications of design and construction*.

一套规范的标准可以控制设计和施工的许多重要技术细节。

【例】We use the water content, *which is the ratio of the weight of water* to that of the solid's.

我们使用的含水量是水的重量与固体土粒重量之比。

【例】Clay minerals have strong surface forces *that are predominant over the gravity forces*.

黏土矿物具有很强的表面力，比重力大得多。

4. 译成状语从句

有时定语从句与主语之间的关系实际上是原因、条件、目的、让步、结果、转折等隐含逻辑关系。因此，英译汉时应以逻辑为基础，将定语从句转译成汉语的状语从句。

【例】Iron is not so strong as steel, which is an alloy of iron with some other elements.

铁的强度不如钢高，因为钢是铁与其他一些元素形成的合金。（原因状语）

【例】An improved design of such a large tower must be achieved, which results in more uniformed temperature distribution in it.

这种大型塔的设计必须改进，以使塔内温度分布更均匀。（目的状语）

【例】Iron, which is not so strong as steel, finds wide application.

虽然铁的强度不如钢，但是它仍有广泛的用途。（让步状语）

Unit Six Building Construction
建筑施工

Warming-up

Task 1: Match the English expressions with their corresponding equivalents.

1. road machinery   A. 压实机械
2. digging machinery   B. 挖掘机械
3. concrete machinery   C. 装修机械
4. hoisting machinery   D. 预应力机械
5. earthmoving machinery   E. 桩工机械
6. compaction machinery   F. 起重机械
7. drilling machinery   G. 筑路机械
8. decorating machinery   H. 铲土运输机械
9. pile driving machinery   I. 钻探机械
10. pre-stressed machinery J. 混凝土机械

Task 2: Study the warning labels and write their Chinese meanings.

1.
CAUTION, DANGER

2.
CAUTION, MECHANICAL INJURY

3.
WARNING, LIFT TRUCKS

4.
NO NEARING

5.
NO RIDING

6.
NO SWITCHING ON

Text A A Brief Introduction to the Building Construction
建筑施工简介

Construction is the translation of a design to a reality. It is important that construction follows the design plan exactly so that the structure will perform as it was intended to and the work will be finished within the required time at the **predicted** cost.

Construction operations are generally **classified** according to specialized fields. The building construction can be divided into the following stages:

- Planning the Construction;
- Preparation of **Site**;
- **Earthmoving**;
- **Foundation Treatment**;
- Steel **Erection**;
- **Concrete Placement**;
- Cleaning of Site.

Planning the Construction

The planning starts with a detailed study of **construction plans** and **specifications**. From this study, a list of all items of work is prepared. These includes individual schedules for **procurement** of materials, equipments and labors. This is the most important stage if the job is to be done effectively and economically. The equipments, labors and materials for each stage of the construction must be provided at the correct time, so careful planning is necessary.

Preparation of Site

"Three connections and one **leveling**" assures that a construction site is connected to water, **electric power supplies** and roads, meanwhile the ground is leveled before a project is begun[1]. All surface structures and the growth from the site should be removed and cleared.

Earthmoving

Earthmoving includes **excavation** and the placement of earth fill. Excavation generally starts with the separate **stripping** of the **organic topsoil**, which is later reused for **landscaping** around the new structure[2]. Excavation may be done by any of several **excavators**, such as **shovels**, **draglines**, **clamshells**, **cranes** and **scrapers**. After placement of the **earth fill**, it is almost always compacted to prevent subsequent **settlement**.

Foundation Treatment

When **subsurface** investigation reveals structural defects in the foundation area to be used for a structure, the foundation must be strengthened. Water passages, **cavities**, **fissures**, **faults** and other defects are filled and strengthened by **grouting**. Grouting consists of **injection** of fluid mixtures under pressure. The fluids subsequently **solidify** in the voids of the **strata**[3]. Most grouting is done with cement and water mixtures, but other mixture ingredients are **asphalt**, cement, clay and **precipitating** chemicals.

Steel Erection

The construction of a steel structure consists of the **assembly** at the site of **mill**-rolled or shop-**fabricated** steel sections. The steel sections may consist of beams, columns or small trusses which are joined together by **riveting**, **bolting** or **welding**[4].

Concrete Placement

Concrete construction consists of several operations: forming, concrete production, placement and **curing**. Concrete is placed by **chuting** directly from the **mix truck**, where possible, or from **buckets** handled by means of cranes or **cableways**, or it can be **pumped** into place by special concrete pumps[5].

Cleaning of Site

When the construction is complete, the site must be completely cleaned and landscaped, and the building **handed over** to managers and maintenance personnel.

New Words

predicted [prɪˈdɪktɪd] adj. 预测的，预期的
classify [ˈklæsɪfaɪ] vt. 分类，归类
site [saɪt] n. 现场
earthmoving [ˈɜːθˌmuːvɪŋ] n. 土方，土方工作
foundation [faʊnˈdeɪʃn] n. 地基，基础
treatment [ˈtriːtmənt] n. 处理
erection [ɪˈrekʃn] n. 架设，安装
specification [ˌspesɪfɪˈkeɪʃn] n. 规格，规范
procurement [prəˈkjʊəmənt] n. 采购，采购信息
level [ˈlevl] vt. 平整，弄平
excavation [ˌekskəˈveɪʃn] n. 挖掘，开挖
stripping [ˈstrɪpɪŋ] n. 清除，脱模
landscape [ˈlændskeɪp] vt. 对……做景观美化，给……做园林美化
excavator [ˈekskəveɪtə(r)] n. 挖掘机
shovel [ˈʃʌvl] n. 铲子，铁锹
dragline [ˈdræɡlaɪn] n. 牵引绳索，拉铲挖土机

clamshell [ˈklæmʃel] n. 抓岩机，抓斗
crane [kreɪn] n. 吊车，起重机
scraper [ˈskreɪpə(r)] n. 铲土机
settlement [ˈsetlmənt] n. 沉降
subsurface [ˈsʌbˌsɜːfəs] adj. 表面下的，地下的
cavity [ˈkævəti] n. 洞穴
fissure [ˈfɪʃə(r)] n. 裂缝，裂隙
fault [fɔːlt] n. 断层
grout [graʊt] vt. 灌浆
injection [ɪnˈdʒekʃn] n. 注入
solidify [səˈlɪdɪfaɪ] vt. & vi. 使凝固，固化
strata [ˈstrɑːtə] n. 地层，岩层 (stratum的名词复数)
asphalt [ˈæsfælt] n. 沥青，柏油
precipitate [prɪˈsɪpɪteɪt] vt. & vi. [化] (使)沉淀
assembly [əˈsembli] n. 装配，组装
mill [mɪl] n. 工厂

fabricate [ˈfæbrɪkeɪt] vt. 制造，装配
truss [trʌs] n. 构件，构架
riveting [ˈrɪvɪtɪŋ] n. 铆接（法）
bolting [bəʊltɪŋ] n. 螺栓连接
welding [weldɪŋ] n. 焊接法，定位焊接

curing [ˈkjʊərɪŋ] n. 固化，养护
chute [ʃuːt] vt. 用斜槽或斜道运送
bucket [ˈbʌkɪt] n. 铲斗
cableway [ˈkeɪbəlˌweɪ] n. 索道，缆道
pump [pʌmp] n. 泵，vt. 用泵抽

Phrases & Expressions

concrete placement 混凝土浇筑
construction plans 建筑图纸
electric power supplies 电力供应
organic topsoil 有机土层

earth fill 填方，填土
mix truck 混凝土搅拌车
hand over 移交，交出

Notes on Text

[1] "Three connections and one leveling" assures that a construction site is connected to water, electric power supplies and roads, and that the ground is leveled before a project is begun.

"三通一平"确保施工工地通水、通电和通路，并在项目开始前平整土地。

[2] Excavation generally starts with the separate stripping of the organic topsoil, which is later reused for landscaping around the new structure.

挖掘过程通常是以清除地表的有机土层开始的，这些土会被再利用来美化新建建筑的环境。

[3] Grouting consists of injection of fluid mixtures under pressure. The fluids subsequently solidify in the voids of the strata.

灌浆是在压力下注入流体的混合物。流体随后在地层的空隙间凝固。

[4] The steel sections may consist of beams, columns or small trusses which are joined together by riveting, bolting or welding.

钢材包括钢梁、钢柱或通过铆钉、螺钉或者焊接连接而成的小型构件。

[5] Concrete is placed by chuting directly from the mix truck, where possible, or from buckets handled by means of cranes or cableways, or it can be pumped into place by special concrete pumps.

在可能的情况下，混凝土可以通过斜槽直接从搅拌车滑下进行浇筑，或者可以从起重机或架空索道控制的铲斗里倒下浇筑，还可以通过特殊混凝土泵浇筑于现场。

Exercises

I. Answer the following questions according to the text.

1. Why is planning the most important part for building construction?
2. What are the main stages of building construction?
3. What does "three connections and one leveling" require?
4. What kinds of excavation facilities are mentioned in the text?
5. How many operations are there for concrete construction? What are they?

II. Match the English words with their Chinese equivalents.

1. classify   A. 泵，用泵抽
2. foundation   B. 对……做景观美化
3. specification   C. 沉降
4. landscape   D. 装配，组装
5. subsurface   E. 挖掘，开挖
6. excavation   F. 规格，规范
7. pump   G. 地基，基础
8. cableway   H. 分类，归类
9. settlement   I. 地下的
10. assembly   J. 索道，缆道

III. Read the following passage and fill in the blanks with the words given in the box.

| environments | causes | eliminated | chemical | movement |
| underlying | tolerably | durable | surface | designer |

Foundation is a part of a structure. It is usually placed below the _____ of the ground, and transmits the load to the _____ soil or rock. All soils compress when they are loaded and cause supported structure to settle. The important requirements in the design of foundations are that the total settlement of the structure shall be limited to a _____ small amount and that differential settlement of various parts of the structure shall be _____ as nearly as possible.

Foundation _____ must consider the effects of construction on buildings nearby, and effects on the _____ of such factors as pile driving vibrations, pumping and discharge of ground water, the disposal of waste materials and the operation of heavy mechanical equipment.

Foundation must be _____ to resist attack by aggressive substances in the sea and river, in soil and rock and in ground water. They must also be designed to resist or to suit _____ from external causes such as seasonal moisture changes in the soil, landslide, earthquake and mining subsidence.

Generally speaking, the strength of soil increases with depth. But it can happen that it becomes weaker with depth. Therefore, in choosing the foundation pressure and level for this will give you an idea of the settlements. There are several _____ of settlement apart from the forces due to load. They are the frost action, _____ change in the soil, underground erosion by following water, reduction of the underground water, the construction tunnels nearby or vibrating machinery such as vehicles.

Ⅳ. **Translate the following English into Chinese or Chinese into English.**

1. electric power supplies
2. steel erection
3. placement of the earth fill
4. fluid mixtures
5. structural defects
6. 混凝土浇筑
7. 建筑图纸
8. 维修人员
9. 地基处理
10. 有机土层

Text B Safety Factors in Design a Building
楼房设计中的安全因素

Design of buildings for both normal and emergency conditions should always incorporate a safety factor against failure. This section presents general design principles for protection of buildings and **occupants** against high winds, earthquakes, water, fire, and lightning.

Wind Protection（Fig.6-1）

For practical design, wind and earthquake may be treated as **horizontal**, or lateral **loads**[1]. Although wind and **seismic loads** may have **vertical components**, these generally are small and readily resisted by columns and **bearing walls**.

In areas where the probability of either a strong earthquake or a high wind is small, it is nevertheless advisable to provide in buildings

Fig.6-1 Wind protection

considerable **resistance** to both types of load. In many cases, such resistance can be incorporated with little or no increase in costs over designs that ignore either high wind or seismic resistance.

Protection Against Earthquakes (Fig.6-2)

Building should be designed to withstand **minor earthquakes** without damage, because they may occur almost everywhere. The principles require that collapse be avoided, **oscillations** of building **damped**, and damage to both structural and nonstructural components minimized[2]. A seismic design of buildings should make allowance for large drift, for example, by providing gaps between **adjoining** building components not required to be rigidly connected together and by permitting sliding of such components. Thus, **partitions** and windows should be free to move in their frames so that no damage will occur when an earthquake wrecks the frames.

Fig.6-2　Protection against earthquakes

Protection Against Water (Fig.6-3)

Whether thrust against and into a building by a flood, driven into the interior by a heavy rain, leaking from pluming, or seeping through the exterior **enclosure**, water can cause costly damage to a building[3].

Protective measures may be divided into two classes: **floodproofing** and **waterproofing**. Floodproofing provides protection against flowing surface water, commonly caused by a river overflowing its banks. Waterproofing provides protection against **penetrating** through the exterior enclosure of buildings of groundwater, rainwater, and melting snow.

Fig.6-3　Protection against water

Protection Against Fire (Fig.6-4)

There are two distinct aspects of fire protection: life safety and property protection. It is not possible to eliminate all **combustible** materials or all potential **ignition sources**. Thus, in most cases, an adequate fire protection plan must assume that unwanted fire will occur despite the best efforts to prevent them. Means must be provided to minimize the losses caused by the fires that do occur.

Fig.6-4　Protection against fire

The first **obligation** of designers is to meet legal requirements while providing the facilities required by the client. In particular, the requirements of the applicable **building code** must be met[4]. The building code will contain fire safety requirement, or it will specify some recognized standard by reference. Many owners will also require that their own insurance carrier be consulted to obtain the most favorable **insurance rate**, if for no other reasons.

Lightning Protection（Fig.6-5）

Lightning, a high-voltage, high-current electrical **discharge** between clouds and the ground, may strike and destroy life and property anywhere thunderstorms have occurred in the past. Buildings and their occupants, however, can be protected against this hazard by installation of a special electrical system. Because an incomplete or poor installation can cause worse damage or injuries than no protection at all, a **lightning protection system** should be designed and installed by experts.

Fig.6-5　Lightning Protection

As an addition to other electrical systems required for a building, a lightning-protection system increases the construction cost of a building. A building owner therefore has to decide whether potential losses justify the added expenditure. In doing so, the owner should compare the cost of insurance to cover losses with the cost of the protection system.

New Words

occupant [ˈɒkjəpənt] n. 占有人；居住者
horizontal [ˌhɒrɪˈzɒntl] adj. 水平的
load [ləʊd] n. 负荷；负担；装载
seismic [ˈsaɪzmɪk] adj. 地震的；由地震引起的
resistance [rɪˈzɪstəns] n. 抵抗；阻力；抗力
damp [dæmp] vi. [物]阻尼；减幅
oscillation [ˌɒsɪˈleɪʃn] n. 振动；波动；<物>振荡
adjoin [əˈdʒɔɪn] vt. & vi. 邻近；附加；接，贴连

partition [pɑːˈtɪʃn] n. 隔离物；隔墙
enclosure [ɪnˈkləʊʒə(r)] n. 圈占；围绕
floodproofing [ˈflʌdˌpruːfɪŋ] 防洪
waterproofing [ˈwɔːtəˌpruːfɪŋ] 防水
penetrate [ˈpenətreɪt] vt. 穿透，刺入；渗入
combustible [kəmˈbʌstəbl] adj. 易燃的，可燃的
obligation [ˌɒblɪˈɡeɪʃn] n. 义务，责任
discharge [dɪsˈtʃɑːdʒ] v. 放出；流出

Phrases & Expressions

seismic load 地震荷载
vertical component 垂直分量
bearing wall 承重墙
minor earthquakes 轻微地震
protective measures 防护措施

ignition sources 火源
building code 建筑规范
insurance rate 保险费率
lightning protection system 防雷系统

Notes on Text

[1] For practical design, wind and earthquake may be treated as horizontal, or lateral loads.
在实际设计中，风和地震可以视为水平荷载或横向荷载。

[2] The principles require that collapse be avoided, oscillations of building damped, and damage to both structural and nonstructural components minimized.
这些原则要求避免倒塌，阻尼建筑物的振动，并尽量减少对结构和非结构部件的损坏。

[3] Whether thrust against and into a building by a flood, driven into the interior by a heavy rain, leaking from pluming, or seeping through the exterior enclosure, water can cause costly damage to a building.
无论是洪水冲撞或冲进建筑物，还是大雨灌入建筑物内部，或是从倾斜处泄漏，或是从外部围栏渗漏，水都会给建筑物造成巨大的损坏。

[4] The first obligation of designers is to meet legal requirements while providing the facilities required by the client. In particular, the requirements of the applicable building code must be met.
设计者的首要义务是在提供客户需要的设施的同时满足法律要求，特别是适用的建筑规范要求。

Exercises

I. Choose the best answers according to the text.

1. The vertical component of wind and seismic loads can be resisted by columns and _____.
   A. balcony   B. bearing walls
   C. stairs   D. windows

2. Building should be designed to withstand _____ without damage, which may occur almost everywhere.
   A. minor earthquakes   B. hurricane

C. mud-rock flow　　　　　　　　D. seaquake

3. _____ will penetrate through the exterior enclosure of buildings and cost damage.

　　A. Groundwater　　　　　　　　　B. Melting snow
　　C. Rainwater　　　　　　　　　　D. All of above

4. Which of the following statement is True according to the passage?

　　A. Partitions and windows should be fixed in their frames so as to avoid damage in earthquake.
　　B. Eliminating all combustible materials or all potential ignition sources is attainable in a building.
　　C. An incomplete or poor installation of electrical system can cause worse damage or injuries than no protection at all in a lightning protection.
　　D. Waterproofing provides protection against flowing surface water, commonly caused by a river overflowing its banks.

5. It is essential to install _____ to protect buildings and their occupant against lightning.

　　A. a special electrical system　　　　B. a regular electrical system
　　C. a special electric system　　　　　D. a regular electric system

II. Fill in the blanks with the following words. Change the forms if necessary.

| resistance | obligation | seismic | horizontal |
| occupant | penetrate | load | combustible |

1. The plumbline is always perpendicular to the _____ plane.
2. The sun's rays can _____ water up to 10 feet.
3. They felt under no _____ to maintain their employees.
4. An efficient bulb may lighten the _____ of power stations.
5. Earthquakes produce two types of _____ waves.
6. Generally speaking there was no _____ to the idea.
7. The present _____ is looking for two females to share the four-bedroom house.
8. Don't smoke near _____ materials.

III. Translate the following English into Chinese or Chinese into English.

1. design principle　　　　　6. 轻微地震
2. seismic load　　　　　　　7. 保护措施
3. vertical component　　　　8. 保险费率

4. fire protection
5. bearing wall
9. 防雷系统
10. 建筑规范

Oral Practices

Conversation One

A: As far as you know, your company has succeeded in winning the tender of Commercial Center. Congratulations!

B: Thank you. It is quite an event in our company.

A: Can you tell me your construction plan of Commercial Center?

B: With pleasure. It covers a building area of 98,600 square meters. It is an important project of the city.

A: I see. Indeed, it is one of the biggest and also the most important works in our city. How long can it be completed in accordance with the contract?

B: It will be completed within eighteen months, I suppose.

A: And when are you going to break ground?

B: It will be start in September this year.

A: When are you going to finish the foundation and the main structure, sir?

B: They are going to be finished at the end of the next year.

A: How about the installation and decoration works?

B: Installation and decoration should be interpenetrated.

A: Your planning is advanced and practical, leaving me a very deep impression. Thank you for your introduction.

B: You are welcome.

Conversation Two

A: What do you do before mixing concrete?

B: We usually prepare the materials—cement, fine aggregates and coarse aggregates.

A: Anything else?

B: Yes. We know that all sands and stones should be hard, sound and clean aggregates, and free from dust, clay and other organic matters.

A: Right. Did you put in the material of concrete aggregate accurately?

B: Certainly. We put in the materials in proportion to weight.

A: What's the grade of concrete used for the foundation?

B: It's 500.

A: Is the mixing time of concrete long enough?

B: Sure. Concrete is being mixed according to the required time.
A: Please show me the pressure testing record of the concrete.
B: No problem. All the test cubes are over there.
A: Let's have a look.
B: OK.

Translation Skills—建筑英语翻译之名词性从句的翻译

1. 名词性从句概述

在句子中起名词作用的句子叫作名词从句（Noun Clauses）。名词从句的功能相当于名词词组，它在复合句中能担任主语、宾语、表语、同位语等，因此根据它在句中不同的语法功能，名词从句又可分别称为主语从句、宾语从句、表语从句和同位语从句。在翻译时，大多数语序可以不变，即可按原文的顺序译成相应的汉语，但有时也需要一些其他处理方法。

引导名词性从句的连接词可分为以下三类：

（1）连接词：that, whether, if（不充当从句的任何成分）。
（2）连接代词：what, whatever, who, whoever, whom, whose, which。
（3）连接副词：when, where, how, why。

2. 建筑英语翻译中名词性从句的翻译方法

（1）主语从句。

构成主语从句的方式有下列两种：

①关联词或从属连词位于句首的从句＋主句谓语＋其他成分。

它们一般译在句首，作为主从复合句的主语。这样的词有关联词what, which, how, why, where, who, whatever, whoever, whenever, wherever及从属连词that, whether, if。

【例】*How this material can be used* depends upon its properties.

这种物质能得到怎么样的利用要由它的性质决定。

【例】*That substances expand when heated and contract when cooled* is a common physical phenomenon.

物体热胀冷缩是普遍的物理现象。

【例】*Whether an increase of the design stress* is warranted depends on a number of conditions, such as type of loading, and frequency and magnitude of overloads.

设计应力的增大是否有保证取决于诸多因素，如载荷的类型、超载的频率和大小。

②It＋谓语＋that（whether）引导的从句。

如果先译主句，可以顺译为无人称从句；有时也可先译从句，再译主句。如果先译从句，便可以在主句前加译，如：

【例】*It is often unclear whether* certain building elements serve a structural function in addition to other functions.

某些建筑构件除有其他功能外是否还发挥结构功能常常是不清楚的。

【例】*It has been found that* some prestressed sleepers do not crack, and disintegration due to the opening and closing of cracks is therefore avoided.

现已发现某些预应力轨枕不开裂，所以避免了因裂缝的张合引起的断裂。

（2）表语从句。

表语从句是位于主句的联系动词后面，充当主句表语的从句，它也是由that, what, why, how, when, where, whether等连词和关联词引导的。一般来讲，可以先译主句，后译从句。

【例】An advantage of prestressed concrete is *that the concrete and the steel are severely tested during the prestressing operation.*

预应力混凝土的一个优点是在施加预应力过程中混凝土和钢材均受到严格考验。

【例】A fundamental premise is *that the basic load carrying mechanism for all structures is the same and is the best examined with reference to concepts of shear and moment.*

一个基本前提是，对于所有结构，基本承载机制是相同的并且根据剪力和弯矩的概念能进行很好的检验。

（3）宾语从句。

一般来说，英语的宾语从句可按原文的语序翻译，但有些介词像except, besides, but, as to 等的宾语从句，应按汉语的习惯译在主句之前。

【例】A crack in the tension zone of the concrete implies *that some slip has occurred between the steel reinforcement and the surrounding concrete.*

在混凝土受拉区出现裂缝意味着钢筋与周围混凝土之间产生了一些滑移。

【例】Engineering metals are used in industry in the form except *that aluminum may be used in the form of a simple metal.*

除了铝可以纯金属形态使用外，各种工程金属都是以合金形式应用于工业的。

【例】Energy can be transmitted from one point to another *by means of what is spoken of as "wave motion."*

利用所谓的波动就可以把能量从一点传递到另一点。

（4）同位语从句。

同位语从句是一种用来进一步说明主句中某一个名词或代词具体内容的句子，通常用从属连词that, whether引导，有时也可用wh-关系副词引导。同位语从句的翻译常采用下列方法。

①翻译时保持原文的语序。

【例】Concrete has the advantage *that it is placed in a plastic condition and is given the desired shape and texture by means of the forms.*

混凝土的优越性就在于浇筑时它处于塑性状态，因此可借助模板获得设计所要求的形状和特征。

②将同位语从句译成定语从句，即译成"……的"。

【例】The internal forces are determined on the assumption *that, in a limit state, the section contains an inclined crack.*

内力是根据极限状态下截面中存在一条斜裂缝的假定来确定的。

③在同位语从句前添加冒号、破折号或添加"即""这""这样"等词。

【例】There is the inherent structural disadvantage *that the core is subjected to the entire weight of the building.*

该结构体系也存在内在缺陷，即核心区承受着建筑物的全部重量。

Unit Seven House Facilities
房屋设备

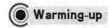 Warming-up

Task 1: Study the internal water supply system chart.

1. valve
2. valve well
3. water meter
4. valve seal
5. water pump
6. water plug
7. inlet pipe
8. outlet pipe
9. water tank
10. shower
11. wash basin
12. water closet pan
13. bathtub
14. tap
15. closet basin
16. branch pipe
17. riser pipe
18. horizontal pipe

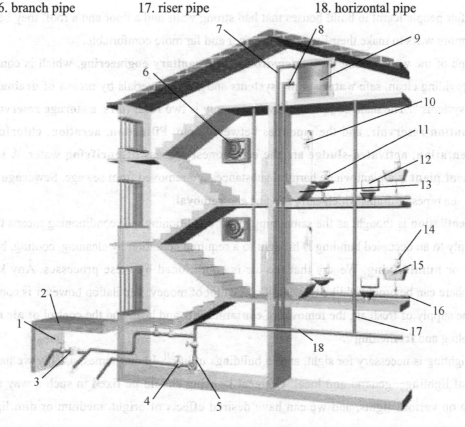

Task 2: Match the English expressions with their corresponding equivalents according to the above chart.

Group A

1. valve　　　　　　A. 水泵
2. valve well　　　　B. 进水管
3. water meter　　　 C. 闸门
4. valve seal　　　　D. 水表
5. water pump　　　 E. 出水管
6. water plug　　　　F. 水箱
7. inlet pipe　　　　 G. 止回阀
8. outlet pipe　　　　H. 消火栓
9. water tank　　　　I. 闸门井

Group B

10. shower　　　　　 A. 水平管
11. wash basin　　　　B. 坐便器
12. water closet pan　 C. 浴盆
13. bathtub　　　　　D. 水龙头
14. tap　　　　　　　E. 洗面盆
15. closet basin　　　 F. 立管
16. branch pipe　　　 G. 洗涤盆
17. riser pipe　　　　 H. 淋浴器
18. horizontal pipe　　 I. 支管

Text A　A Brief Introduction to the House Facilities
房屋设备简介

After people learnt to build houses that had strong walls and a floor and a roof, they began to invent more ways to make their living much easier and far more comfortable.

One of the ways among them is **environmental sanitary engineering**, which is concerned with providing clean, safe **water supply systems** and waste materials **by means of drainage** and **sewer** systems[1]. The main equipment in water supply is two **reservoirs**, a **storage reservoir** and **distribution reservoir**, and the pipelines between them. **Filtration, aeration, chlorination, sedimentation, activated-sludge** are the commonest process of **purifying** water. **A sewage treatment plant** is a plant where harmful substances are removed from sewage. **Sewerage** means mainly the pipes or drains which carry sewage and **removal**.

Ventilation is thought as the same thing as air conditioning. Air conditioning means that the air supply to an occupied building is brought to a required condition by cleaning, cooling, heating, drying or **humidifying**. We say that the air is conditioned by these processes. Any kind of atmosphere can be produced like this, but it costs a lot of money. Ventilation however is concerned with the supply of fresh air, the removal of **contaminants** and heat, and the control of **air motion** for cooling and **freshening**[2].

Lighting is necessary for sight, and in buildings utility[3]. In our homes, usually we have two kinds of lighting—general and local. **General lighting** should be fixed in such a way that by putting on various lights, and we can have desired effects of bright, medium or dim light for various occasions. **Local lighting** is produced by floor or table lamps or from light tubes.

Families in modern cities are not often seem need supplies of either coal or wood for cooking and heating but gas. **Natural gas** is found underground and under the sea. It is brought to the surface and then taken to a purifying plant, then either piped directly to houses or made into liquid gas and taken to other towns.

The elevator is a convenient labor saving device **installed** in tall buildings. With the help of an elevator, people and material can be moved up or down a building with great efficiency and safety[4]. Today, elevators are **computerized**.

What's more? That's the development of appliances in home, such as TV sets, telephones, computers and video recorders etc, which give people a **nondescript** convenience in communication, work and recreation. At the moment, the "smart house" created by the Building Research Association, whose many functions are controlled by computers, is already being used in the United States of America[5]. Let's think about it. Pick up your cellphone in your car, dial home and simply by pushing the numbers on your phone, you tell your house what to do ……

Is it the beginning or the end of home automation?

New Words

facility [fəˈsɪləti] n. 设备
sanitary [ˈsænətri] adj. 卫生的，清洁的
drainage [ˈdreɪnɪdʒ] n. 排水，放水
sewer [ˈsuːə(r)] n. 污水管，下水道
reservoir [ˈrezəvwɑː(r)] n. 蓄水池，水库
filtration [fɪlˈtreɪʃn] n. 过滤
aeration [eəˈreɪʃn] n. 曝气
chlorination [ˌklɔːrɪˈneɪʃn] n. 氯化
sedimentation [ˌsedɪmenˈteɪʃn] n. 沉淀，沉降
activated-sludge [ˈæktɪveɪtɪd-slʌdʒ] n. 活性污泥
purify [ˈpjʊərɪfaɪ] vt. 净化
sewage [ˈsuːɪdʒ] n. 污水，废水

sewerage [ˈsuːərɪdʒ] n. 污水工程，污物处理（系统）
removal [rɪˈmuːvl] n. 清除
ventilation [ˌventɪˈleɪʃn] n. 通风
humidify [hjuːˈmɪdəfaɪ] vt. 使湿润，使潮湿
contaminant [kənˈtæmɪnənt] n. 污染物，致污物
freshen [ˈfreʃn] vt. 使新鲜，使清爽
install [ɪnˈstɔːl] vt. 安装
computerize [kəmˈpjuːtəraɪz] vt. （使）电子计算机化
nondescript [ˈnɒndɪskrɪpt] adj. 难以描述的

Phrases & Expressions

environmental sanitary engineering 环境卫生工程
water supply systems 给水系统
by means of 依靠，凭借

storage reservoir 蓄水库
distribution reservoir 配水库
sewage treatment plant 污水处理厂
air motion 空气流动

general lighting 总体采光
local lighting 局部采光
natural gas 天然气

Notes on Text

[1] One of the ways among them is environmental sanitary engineering, which is concerned with providing clean, safe water supply systems and waste materials by means of drainage and sewer systems.

其中的一个方法就是环境卫生工程，即提供清洁、安全的给水系统并涉及利用排水和下水道系统处置剩余水和废料。

[2] Ventilation however is concerned with the supply of fresh air, the removal of contaminants and heat, and the control of air motion for cooling and freshening.

通风是指供给新鲜空气，排除污浊气体和热气，以及为使空气冷却和清新而调节空气的流量。

[3] Lighting is necessary for sight, and in buildings utility.

照明（采光）在建筑物中对视觉和实用性都是必需的。

[4] With the help of an elevator, people and material can be moved up or down a building with great efficiency and safety.

借助于电梯，人和物可以高效率地、安全地在大楼内上下。

[5] At the moment, the "smart house" created by the Building Research Association, whose many functions are controlled by computers, is already being used in the United States of America.

目前，由建筑研究协会研制的"精巧屋"已在美国得到应用，其很多功能都是由计算机控制的。

Exercises

I. **Answer the following questions according to the text.**

1. What is the main equipment in water supply?
2. What's the difference between ventilation and air conditioning?
3. What are the two kinds of lighting?
4. How can natural gas be taken to houses and towns?
5. What is the function of an elevator nowadays?

Ⅱ. Match the English words with their Chinese equivalents.

1. facility   A. 清除
2. humidify   B. 卫生的
3. nondescript   C. 设备
4. install   D. 水库
5. computerize   E. 净化
6. reservoir   F. 使湿润
7. purify   G. 使新鲜
8. freshen   H. 难以描述的
9. sanitary   I. 安装
10. removal   J. （使）计算机化

Ⅲ. Read the following passage and fill in the blanks with the words given in the box.

| over | light-tube | indirect | dim | separate |
| general | visual | local | occasions | functional |

Lighting is one of the basic concepts of interior decoration. There are unlimited possibilities to produce many _____ effects with the light.

The light fixtures should be _____ as well as decorative. In our homes, usually we have two kinds of lighting — _____ and local.

General lighting should be designed in such a way that by turning on various lights, and we can have desired effects of bright, medium or _____ light for various _____.

Too many fluorescent lights produce a monotonous effect. We should go for a mixture of direct and _____ lighting arrangement.

In the living rooms, concealed lighting is more effective. But a pair of identical lamps may stand at each end of a sofa on small tables. Kitchen should be particularly well lighted. One should have wall and ceiling light _____ the working places in it. Bedroom lights should usually be _____ lights. Lights attached to either side of a dressing table mirror or one _____ placed at the top of the mirror would be an excellent idea. In the study, a desk should have a _____ lamp.

Ⅳ. Translate the following English into Chinese or Chinese into English.

1. purifying plant   6. 房屋设备
2. liquid gas   7. 环境卫生工程
3. harmful substances   8. 下水道系统
4. fresh air   9. 空气流动
5. light tube   10. 局部采光

Text B Tongling Recluse
铜陵山居

Tradition and modern both **prevail** in our contemporary Chinese architectural practice. Tongling **Recluse**（Fig.7-1）locates in a **secluded** village in southern Anhui, and the project originates from a normal house **integrating** the Hui style with the Chinese riverside style. The existing building occupies a small area of the highest peak in the village. It has not been **inhabited** for more than ten years, thus the present situation is very **dilapidated** with weeds and shrubs **springing up** the surrounding areas. The house has three span in east-west direction and one **span** in south-north direction, and the original roofs and walls are severely damaged.

Fig.7-1 Design concept

Hence, we extend one span straight to the rocky mountain in west direction, add one span in south-north direction forming a **spacious** space for living room. We introduce **curves** in part of the plan, abstract the volume from the original projected outline, and **dissimilate** the original volume into a **transitional** enclosure[1]. **Due to** the damaged conditions and the limited height, we add one more floor to the existing building, **stretch** the twin volumes in space which is lofting from the curves in plan, **head down** the **ridge** at the front part shaping a both scattered and continuous surface in **spatial**. Traditional folded roof blends with the abstracted curves into a whole, **coincide with** the **cosmology** of Chinese culture "Tao **begets** one, One begets two." When looking down from above, the whole roof covered with grey tiles generates a unique shape（Fig.7-2）. Tongling Recluse can not only **merge into** the ancient village but also **stand out** with its special appearance, as well as presenting the interior characteristics from the exterior.

Fig.7-2 Unique shape of the roof

Therefore the house forms a four span space in east-west direction. Starting from the east, the first span is an **anteroom** for living room situating at the south part of the site[2]. Under the second span, the remain broken wall forms a courtyard and a glassy sightseeing platform on the second floor. The bedroom space **accommodating** under the third span is formed with the existing walls on the west[3]. The last span which is straight to the mountain on the west forms a Chinese traditional underoof-space, as well as blends parts of the mountain landscape（Fig.7-3）[4]. Seen from the south **elevation**, the house seems like a contemporary **collage** under a series of traditional context.

Fig.7-3　Underoof-Space and the mountain landscape

We add one more ridge next to the original one on the east elevation that suggest a relationship of twin-buildings, bringing a balance of peace and quiet. The new ridge itself becomes a span with half of it merging into the outline of existing building and the other half becoming a **veranda** under the **eaves**[5]. It creates a contrast between nothingness and **essence** on the east elevation as well as communicates with the landscape of **sunken** courtyard on the south. Different from the（typical Chinese paintings'）**perspectives** scattered horizontally on the southern facade, the limited views through the eastern glass wall lift over a small cliff are more like one-point perspective views[6]. The interior brass decorations, the cornice along with the **suspend** terrace create a feeling of **dynamic** and anticipation（Fig.7-4）.

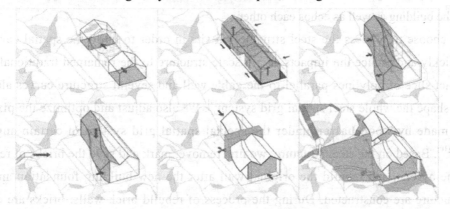

Fig.7-4　Original building modification

The major internal active space which **apposes** to the external corridor space is **staggered** by **a series of** remaining and new walls[7]. Thus the remaining broken wall is used as a sequence of interior cut-off walls. Terrace, living room, dining room, kitchen, courtyard and bedroom are lined up from east to west with privacy decreasing. Stepping into the living room through the external corridor space from the southern stairs, inhabitants could have family gatherings and activities. The first floor is designed to ensure the public view is **unencumbered** so that you can experience the area showcasing the **parlors** original height. On the right, a semi-outdoor terrace offers a magnificent view of the continuous mountains and the entire village. Turning left, inhabitants enter a more private dining space and catering space on the opposite side. There is a bedroom and a bathroom on the first floor and the two more bedrooms and bathrooms （Fig.7-5）in the loft, which can be accessed from the staircase in the living room.

Fig.7-5 Bathroom

Atrium is the core of the floor plan （Fig.7-6）, turning the existing interior space into interior space of the new building. Atrium, together with the west grey space under the eaves, creates a repeatedly transferring relation between emptiness and solidity in the east-west direction[8]. With the suspended stairs and full height glazing on

Fig.7-6 The floor plan

the east,it also achieves a high level of transparency which breaks the limit of traditional house. In the south-north direction, a sequence of **clerestories** separate the walls and **hyperbola** roof, **intensifying** the contemporary sense and the floating ability of the roof. **A string of fenestellas**, with some important part of the building such as bedrooms,stairs, etc, create an opposite scenery in and out the building as well as echos each other.

We choose to process the steel structure **on site** in order to build the spatial curve roof more quickly and reduce the impacts of big scale structure to the remained traditional house. We extract slice of polylines parallel to the **gable** wall and several structure curves along the ridge to shape the whole steel spatial grid system[9]. We also adjust and optimize the **pixel** roof surface made by wood **batten** under the special spatial grid system in certain angle and position[10]. Based on the steel structure, we first remove, mark and save the bricks of remained walls one by one, and rebuild the original wall after the new building foundation and main steel structure are constructed. During the process of rebuild brick walls, bricks are cut into pieces of slices as outside decoration in order to hide the steel structures inside the traditional elements. New walls are composed with different old bricks from local buildings, and we brush the newly part with white paint which is commonly used in Hui style. At the connecting part, we intentionally use antique and finish to create a sense of harmony and unity. Same to the brick walls, columns and roof are built by the old tiles recycled from the old local houses. We invite local craftsmen to do the **masonry** with local technics echoing the local culture on the **tectonic** side as well as showing **sustainable ecological** philosophy.

New Words

prevail [prɪˈveɪl] vi. 流行，盛行；获胜，占优势；说服，劝说
recluse [rɪˈkluːs] n. 隐居者，遁世者，隐士
secluded [sɪˈkluːdɪd] adj. 与世隔绝的；隐退的；偏僻的
integrate [ˈɪntɪɡreɪt] v. 合并；成为一体；加入；融入群体
inhabit [ɪnˈhæbɪt] vt. 居住；在……出现；填满 vi. 居住
dilapidated [dɪˈlæpɪdeɪtɪd] adj. 残破的，衰败的；破旧
span [spæn] n. 跨度，墩距；一段时间
spacious [ˈspeɪʃəs] adj. 宽敞的；广阔的
curve [kɜːv] n. 弧线，曲线 adj. 弯曲的 vt. 使弯曲；使成曲线
dissimilate [ˈdɪsɪmɪleɪt] v. （使）变得不同
transitional [trænˈzɪʃənl、trænˈsɪʃənl] adj. 变迁的，过渡期的；转变的；转移的
stretch [stretʃ] v. 伸展；延伸；持续；n. 伸展；弹性；adj. 可伸缩的；弹性的
ridge [rɪdʒ] n. 背脊，峰；山脊 vt. 使成脊状，使隆起 vi. 使成脊状
spatial [ˈspeɪʃl] adj. 空间的；存在于空间的；受空间条件限制的
cosmology [kɒzˈmɒlədʒi] n. 宇宙学
beget [bɪˈɡet] vt. 产生，引起
anteroom [ˈæntɪruːm] n. 接待室，前厅
accommodate [əˈkɒmədeɪt] vt. 容纳；使适应；向…提供住处
elevation [ˌelɪˈveɪʃn] n. [建]正视图，立视图；高地，高度
collage [ˈkɒlɑːʒ] n. 拼贴画；拼贴艺术；vt. 拼贴；vi. 制作拼贴

veranda [vəˈrændə] n. 阳台，走廊
eave [iːv] n. 屋檐
essence [ˈesns] n. 本质，实质；精华，精髓
sunken [ˈsʌŋkən] adj. 沉没的；凹陷的，下陷的
perspective [pəˈspektɪv] n. 远景，景色；观点，看法；洞察力
suspend [səˈspend] v. 暂停；延缓；悬挂
dynamic [daɪˈnæmɪk] adj. 动态的；动力的；精力充沛的
appose [əˈpoʊz] v. 并列，放……在对面
stagger [ˈstæɡə(r)] vi. 蹒跚；犹豫；动摇；vt. 使蹒跚，使摇摆
unencumbered [ˌʌnɪnˈkʌmbəd] adj. 没有阻碍的，不受妨碍的；无负担的
parlor [ˈpɑːlə(r)] n. 客厅；起居室；（旅馆中的）休息室
atrium [ˈeɪtriəm] n. （现代建筑物开阔的）中庭，天井；心房
clerestory [ˈklɪəstɔːri] n. 天窗，通风窗
hyperbola [haɪˈpɜːbələ] n. 双曲线
intensify [ɪnˈtensɪfaɪ] vt. & vi. （使）增强，（使）加剧
fenestella [ˌfenəˈstelə] n. 小窗，窗状壁龛
gable [ˈɡeɪbl] wall 山墙，山墙墙身，前脸墙
pixel [ˈpɪksl] n. （显示器或电视机图像的）像素
batten [ˈbætn] n. 板条；压条；vt. 用板条或压条固定
masonry [ˈmeɪsənri] n. 石工工程，砖瓦工工程；砖石建筑
tectonic [tekˈtɒnɪk] adj. 构造的，建筑的
sustainable [səˈsteɪnəbl] adj. 可持续的；可支撑的
ecological [ˌiːkəˈlɒdʒɪkl] adj. 生态（学）的

Phrases & Expressions

spring up 跳起；跃起；迅速成长
stand out 突出；坚持；超群；向前跨步
a series of 一系列；一连串
a string of 一系列；一批；一连串
coincide with 与……一致

Notes on Text

[1] We introduce curves in part of the plan, abstract the volume from the original projected outline, and dissimilate the original volume into a transitional enclosure.

在平面上部分引入曲线，一个异化的虚体从原有投影轮廓中抽离出来，使原有体积和转变的围墙不同。

[2] Starting from the east, the first span is an anteroom for living room situating at the south part of the site.

从东起，一跨形成地块南向并作为起居室的前厅空间。

[3] The bedroom space accommodating under the third span is formed with the existing walls on the west.

一跨与西侧的原建筑墙体形成了卧室空间。

[4] The last span which is straight to the mountain on the west forms a Chinese traditional underoof-space, as well as blends parts of the mountain landscape.

增加的一跨向西将山体和营造的部分景观空间纳入其半开放的檐下虚空间。

[5] The new ridge itself becomes a span with half of it merging into the outline of existing building and the other half becoming a veranda under the eaves.

新脊自成一跨，其中一半并入原建筑外轮廓，另一半形成檐下外廊。

[6] Different from the（typical Chinese paintings'）perspectives scattered horizontally on the southern facade, the limited views through the eastern glass wall lift over a small cliff are more like one-point perspective views.

与南立面的横向展开的散点透视关系不同，东向悬崖方向的姿态更加倾向于一种单点透视的主体画面感。

[7] The major internal active space which apposes to the external corridor space is staggered by a series of remaining and new walls.

与外部廊道空间并置的，即为建筑的主要内部活动空间，由上述一系列新老墙体交错而成。

[8] Atrium, together with the west grey space under the eaves, creates a repeatedly transferring relation between emptiness and solidity in the east-west direction.

整个中庭空间与西侧的檐下景观虚空间形成了东西一线的虚实关系的反复转换。

[9] We extract slice of polylines parallel to the gable wall and several structure curves

along the ridge to shape the whole steel spatial grid system.

我们提取出平行于山墙面方向的若干"切片"式折线以及沿屋脊方向的几条结构曲线，形成整个屋面的钢结构空间网格体系。

[10] We also adjust and optimize the pixel roof surface made by wood batten under the special spatial grid system in certain angle and position.

在这个结构体系下，屋顶的曲面形态可以通过其表面上的像素式的木挂瓦工艺在空间网格下的角度和位置进行更细微的调整和优化。

Exercises

I. Choose the best answers according to the text.

1. What is the style of this project?
   A. The Hui style.   B. Not mentioned.
   C. The Chinese riverside style. D. A mixture of two styles.

2. What did the designer do with the existing building?
   A. Add one more wall.   B. Add one more floor.
   C. Add one more room.   D. Add one more house.

3. How many spans did the house have in east-west direction?
   A. Two.   B. Three.   C. Four.   D. Five.

4. The designer add one more ridge in order to_____.
   A. get more space   B. make it like a building
   C. make it more complex   D. get it a twin building

5. What function does the remaining broken wall have?
   A. Supporting walls.   B. Interior cut-off walls.
   C. Divide the rooms.   D. Keep off the outside.

6. Where is the atrium now?
   A. Its original place.   B. Inside the house.
   C. Outside the house.   D. Not mentioned.

7. How does the designer deal with those old bricks?
   A. Throw them away.   B. Sell them.
   C. Reuse them.   D. Give them away.

II. Fill in the blanks with the given words. Change the form if necessary.

relic	span	spatial	tectonic
batten	masonry	gable	eave

1. A _____ is a long strip of wood that is fixed to something to strengthen it or to hold it firm.

2. It has a dormer roof joining both _____ ends.

3. The batteries had a life _____ of six hours.

4. Changes taking place in the _____ distribution of the population.

5. She was injured by falling _____.

6. These _____ tiles are carved with many animal pictures.

7. The building stands as the last remaining _____ of the town's cotton industry.

8. Most mountains are formed at _____ plate boundaries.

Ⅲ. **Translate the following English into Chinese or Chinese into English.**

1. hyperbola roof
2. the Chinese riverside style
3. sightseeing platform
4. underoof-space
5. the existing wall

6. 落地玻璃
7. 外部廊道
8. 木挂瓦
9. 钢结构
10. 飘板楼梯

Oral Practices

Conversation One

A: Hi, Jack. May I ask you some questions about the heating works?

B: Certainly.

A: What do you do before laying heating pipes?

B: We prepare pipes, fittings and radiators according to the drawings first, and then spray or brush anti-rust pain and aluminum paint on them.

A: Right. What requirements do you follow for rise pipes and mains?

B: Mains must have a fall, and rise pipes must be vertical.

A: I see. Thanks a lot.

B: My pleasure.

Conversation Two

A: Excuse me, sir, what are you doing?

B: Making a square vent pipe.

A: What kind of material, sheet or plate do you use?

B: I usually use sheet.

A: Does the workshop adopt natural draft ventilation or artificial ventilation?

B: Artificial ventilation.

A: What kind of remover does that workshop use?

B: It's a dust-bag remover.

A: That's good. I'd like to know if the air-filter is fixed on the ventilation system.

B: OK, as it is shown on the drawing.

A: What function does it have?

B: To ensure the infusion of fresh air in accordance with the demands.

A: I see. The exhaust pipe of our workshop needs an exhaust cover. Could you please make one for me?

B: Certainly.

Translation Skills—建筑英语翻译之状语从句的翻译

1. 状语从句概述

用来修饰主句或主句的谓语的英语句子叫作状语从句。状语从句一般可分为八大类，分别表示时间、地点、原因、目的、条件、让步、结果和方式。尽管种类较多，但由于状语从句与汉语的结构和用法相似，所以理解和掌握它并不难。状语从句的关键，是要掌握引导不同状语从句的常用连接词和特殊的连接词点。

2. 建筑英语中状语从句的翻译方法

（1）时间状语从句及其译法。

时间状语从句的翻译，主要在于掌握好各种时间意义的连词。时间状语从句在英语句子中的位置相对灵活，但汉译时，有时候就要注意它们的位置。汉语习惯是先发生的事情先讲，表示时间的从句汉译时要提前。当时间顺序很明显时，有时还可以省略关系副词。

【例】Pre-tensioning is a method of prestressing in which the steel tendons are tensioned *before the concrete has been placed in the moulds*.

先张法是一种在模板内浇筑混凝土之前，将钢筋束拉张而施加预应力的方法。

【例】On the site *when further information becomes available*, the engineer can make changes in his sections and layout, but the drawing office work will not have been lost.

在现场若能取得更确切的资料，工程师就可以修改他所做的断面图和设计图，但是绘图室的工作并非徒劳无功。

（2）地点状语从句及其译法。

引导地点状语从句的多为where和wherever。当where引导的从句位于句首时，是一种加强语气的说法，而且含有条件意味，可翻译成"哪里……哪里……"。wherever引导的从句位于句首时，除了有强调的意味外，还有"让步"的意义。可翻译成"不论到哪里；哪里都……"。

【例】*Where there is design*, there is innovation.

哪里有设计，哪里就有创新。

【例】*Wherever foundations are needed*, reinforced concrete for structures will be indispensable.

凡是需要基础的地方，钢筋混凝土结构就必不可少。

（3）原因状语从句及其译法。

引导原因状语从句的从属连词有because，since，as。because表示直接的因果关系，可用于强调句中；since表示的原因语气较弱，往往是指已经知道的原因，可用于省略句中；as的语气更弱，通常表示显而易见的原因。它们引导的原因状语从句可翻译成汉语中的表示"因"的分句。

【例】*Since no material is perfectly rigid*, the imposition of a load always produces a deformation.

没有一种材料是完全刚性的，所以受到荷载作用时材料总会产生变形。

【例】*As air has weight*, it exerts force on any object immersed in it.

空气具有重量，所以它对浸入在它里面的任何物体会施加一个力。

在某些情况下也可翻译成汉语中的因果偏正复句的主句。

【例】*Because energy can be changed from one form into another*, electricity can be changed into heat energy, mechanical energy, light energy, etc.

能量能从一种形式转换为另一种形式，所以电能可以转变为热能、机械能、光能等。

（4）目的状语从句及其译法。

英语的目的状语从句通常位于句末，汉译时可译成后置分句。但是，汉语中表示"目的"的分句常用"为了"作为关联词置于句首，往往具有强调的含义。

【例】A rocket must attain a speed of about five miles per hour *so that it may put a satellite in orbit*.

要把卫星送入轨道，火箭必须得达到每秒大约5英里的速度。

（5）条件状语从句及其译法。

英语中if（如果），unless（如果……不……）引出一般性条件从句，表示某一件事情发生或一种现象的出现，在某种程度上依赖另一正在发生的事情或现象。连词 unless 相当于 if...not...

【例】Any body above the earth will fall *unless it is supported by an upward force equal to its weight*.

地球上的任何物体都会落下来，除非它受到一个与其重量相等的力的支持。

但是根据对原文逻辑含义的理解或出于汉语习惯的考虑，有时由where或when引导的地点状语从句也会翻译成条件从句。

【例】*Where the available rolled beams are not sufficient to carry the load*, plate girders have to be used.

如果现存的轧制梁不足以承担荷载，就必须使用板梁。

【例】*When the reinforcement is strongly bonded to the concrete*, a strong, stiff, and ductile construction material is produced.

如果钢筋与混凝土牢牢粘连在一起，就产生出一种坚韧的具有延性的建筑材料。

（6）让步状语从句及其译法。

英语中的让步状语从句可由although，though，as，even though，even if等引导。

【例】*Although aluminum alone is not particular strong*, it can form a very strong light alloy.

虽然纯铝的强度不特别高，但它能够形成高强度的轻合金。

（7）结果状语从句及其译法。

英语和汉语都将表示结果的从句放在主句之后，翻译时可采用顺译法，但应注意不能拘泥于引导结果状语从句的连词 so...that, such...that 等词义而一概翻译成"因而""结果""如此……以至于……"等。翻译时应尽量避开连词，以免使译文过于欧化。

【例】Electricity is *such an important energy that* modern industry couldn't develop without it.

电是非常重要的一种能量，没有它，现代化工业就不能发展。

Unit Eight Interior Decoration
室内装饰

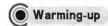

Task 1: Match the English expressions with their corresponding equivalents.

1. Chinese classical decoration style   A. 巴洛克装饰风格
2. Japanese decoration style   B. 地中海式装饰风格
3. Rural adornment style   C. 日式装饰风格
4. Boreal Europe adornment style   D. 中式古典装饰风格
5. Mediterranean adornment style   E. 田园装饰风格
6. Baroque adornment style   F. 北欧装饰风格

Task 2: Match the pictures with the above decoration styles in Task 1.

1. _____ 2. _____ 3. _____

4. _____ 5. _____ 6. _____

Text A Green Interior Decoration
绿色室内装饰

Fig.8-1 Green interior decoration

As is known, green **interior decoration** (Fig.8-1) is a fresh, **dynamic** and developing **concept**, which will bring an ideal mode for **sustainable development** to seek the harmonious **coexistence** between human and nature by **science and technology**[1]. It tries to reduce the waste pollution caused by the material waste, and **make** full **use of** natural light and ventilation, and at the same time, through the reasonable layout of green plants, it can purify the indoor environment and improve **air quality**.

In the choice of interior decoration materials, according to the type of architecture, grades and specific requirements of the site and use, we ought to make clever use of the simple sense of material, **linearity** and color to make building decoration meet a certain function, adapt to a certain environment and reflect the best **adornment** effect[2].

Metope Material (Fig.8-2)

We **tend to** choose some environmental wallpaper. Now green design and decoration are trying to return to nature, it is not impossible to select and apply the natural plant adornment metope, for

Fig.8-2 Metope material

example, cotton, **linen** and **cany** goods can be chosen as base materials of natural wall decoration.

Ground Material (Fig.8-3)

Stone, wood, carpet, **ceramic** tile and plastic floor are the five most common and **indispensable** materials on the ground. The **emergence** of new **plastic** floor at present is gradually close to the **environmental protection**. The surface of plastic floor is bright and clean, against resistance, with high strength and light weight. Its color is rich, its construction is convenient, and the cost is lower[3]. Microcrystal is a kind of natural inorganic materials which is also accepted by the public as well.

Fig.8-3　Ground material

Natural Light Source (Fig.8-4)

Interior green decoration is also just based on the use of natural light source through the change of the position of the windows, and if needed, through the **transparent** roof more good daylight can be provided. The full use of natural light source can avoid unnecessary waste of resources, save energy and increase the degree of light.

Fig.8-4　Natural light source

Healthy Color Collocation (Fig.8-5)

Color applied properly in decoration is not only the comfortable vision, but also a **subtle psychological** influence[4]. Green space needs this color **collocation** of the human nature. The right color in living space can not only benefit the health of the owner, but also make the physical and mental relaxation.

Distribution of Space Function Combination (Fig.8-6)

Reasonable space organization arrangement plays an important part in green interior decoration. The so-called reasonable **distribution**, **in addition to** satisfying the human body and perfecting the outside space, also considers the space functional organization. The fundamental concern of design is

Fig.8-5　Healthy color collocation

how the space can be used. As architects and interior designers create and modify spaces, they should communicate concepts and feelings to all those who see, use and occupy those spaces[5].

Fig.8-6 Distribution of space function combination

New Words

decoration [ˌdekəˈreɪʃn] *n.* 装饰，装潢
dynamic [daɪˈnæmɪk] *adj.* 动态的
concept [ˈkɒnsept] *n.* 观念，概念
coexistence [ˌkəʊɪɡˈzɪstəns] *n.* 共存
linearity [ˌlɪnɪˈærəti] *n.* 线型
adornment [əˈdɔːnmənt] *n.* 装饰，装饰品
metope [ˈmetəʊp] *n.* 墙面
linen [ˈlɪnɪn] *adj.* 亚麻的，亚麻制品的
cany [ˈkeɪni] *adj.* 藤的，藤制的
ceramic [səˈræmɪk] *adj.* 陶瓷的

indispensable [ˌɪndɪˈspensəbl] *adj.* 不可缺少的
emergence [iˈmɜːdʒəns] *n.* 出现，发生
plastic [ˈplæstɪk] *adj.* 塑料的
transparent [trænsˈpærənt] *adj.* 透明的
subtle [ˈsʌtl] *adj.* 微妙的，精细的
psychological [ˌsaɪkəˈlɒdʒɪkl] *adj.* 心理的，心理学的
collocation [ˌkɒləˈkeɪʃn] *n.* 搭配，配置
combination [ˌkɒmbɪˈneɪʃn] *n.* 组合，结合
distribution [ˌdɪstrɪˈbjuːʃn] *n.* 分布，分配

Phrases & Expressions

interior decoration 室内装饰
as is known 众所周知
sustainable development 可持续发展
science and technology 科技
make use of 利用

air quality 空气质量
tend to 倾向于
environmental protection 环保
in addition to 除了

Notes on Text

[1] As is known, green interior decoration is a fresh, dynamic and developing concept, which will bring an ideal mode for sustainable development to seek the harmonious coexistence between human and nature by science and technology.

众所周知，绿色室内装饰是一个新兴的、动态的和发展中的概念，它通过科技手段带来一种可持续发展的，实现人与自然和谐共存的理想模式。

[2] In the choice of interior decoration materials, according to the type of architecture, grades and specific requirements of the site and use, we ought to make clever use of the simple sense of material, linearity and color to make building decoration meet a certain function, adapt to a certain environment and reflect the best adornment effect.

在选择室内装饰材料时，需要根据建筑的类型、档次和使用部位的具体要求，来巧妙合理地运用材料的质感、线型和色彩，以便使建筑装饰满足一定的功能，适应一定的环境，发挥出最佳的装饰效果。

[3] The surface of Plastic floor is bright and clean, against resistance, with high strength and light weight. Its color is rich, its construction is convenient, and the cost is lower.

塑料地板，表面光洁，耐老化性好，轻质高强，颜色丰富，施工方便，造价相对较低。

[4] Color applied properly in decoration is not only the comfortable vision, but also a subtle psychological influence.

色彩运用得当不仅使视觉上感觉舒服，心理上也有一种潜移默化的影响。

[5] As architects and interior designers create and modify spaces, they should communicate concepts and feeling to all those who see, use and occupy those spaces.

在建筑师和室内设计师创造和修改空间格局时，他们要与那些看到、使用和占有这些空间的人们交流理念和感觉。

Exercises

I. **Answer the following questions according to the text.**

1. What will we consider when choosing interior decoration materials?
2. What can be chosen as the basic materials of natural wall decoration?
3. What's the advantage of making use of natural light source?
4. What kind of ground materials is the most environmental?
5. How to understand the word "reasonable" in the last paragraph?

Ⅱ. Match the English words with their Chinese equivalents.

1. decoration   A. 不可缺少的
2. indispensable   B. 出现，发生
3. coexistence   C. 微妙的，精细的
4. plastic   D. 概念，观念
5. subtle   E. 共存
6. transparent   F. 动态的
7. emergence   G. 塑料的
8. concept   H. 墙面
9. metope   I. 装饰，装潢
10. dynamic   J. 透明的

Ⅲ. Read the following passage and fill in the blanks with the words given in the box.

| sculpture | get bored of | marble | what | odd |
| incorporate | in abundance | evolved from | distinct | architecture |

The _____ of this time was called Baroque because it was considered to be very _____. This architecture _____ Renaissance architecture in Italy in the 1600s when the architects there began to _____ the same symmetry and forms that they had been using for the past 200 years. When this happened, they began to make curving facades and used the double curve on many different buildings. _____, gilt and bronze were the materials the Baroque architects used _____. Oval was the most _____ shape of the Baroque style and was a very common shape among the Baroque buildings. Gian Lorenzo Bernini and Francesco Borromini were the two main architects of the Baroque era. Since Bernini's first medium was _____, this was _____ he liked to _____ into his buildings. Francesco Borromini was a sculpture and mason who incorporated many shapes and different forms into his designs.

Ⅳ. Translate the following English into Chinese or Chinese into English.

1. interior decoration   6. 墙面材料
2. environmental protection   7. 自然光源
3. air quality   8. 色彩搭配
4. science and technology   9. 地面材料
5. sustainable development   10. 空间组合

Text B Five Elements of Interior Design in Decorating
室内装饰设计的三个基本因素

Design is the art of combining different elements in a pleasing way. The five basic elements of design useful in **decorating** a house are color, **texture**, line, form and space. Each element has its unique characteristics, but once put and mixed together an attractive and creative outcome is produced, thus beautifying the home[1]. It is important to understand these elements and how they work together to create the right style.

Color (Fig.8-7)

When you select wallpaper, paint, window hangings, furniture and even carpet, one of the first things you should take into consideration is color. Color sets the tone of a room. It can make it feel bigger or smaller, warmer or cooler. Your color **palette** will give your space a certain mood that visitors will notice instantly. One of the most important considerations in **interior design** is selecting

Fig.8-7 Color

colors that work well together. Color can be used to draw attention to your focal point, or **disguise** your least favorite features. When selecting colors, try to pick several colors in the same family, or that **compliment** each other. Don't try to get several items to match exactly. It is nearly impossible to match, it's much easier to select different shades that can work together.

Texture (Fig.8-8)

The next thing you may notice when selecting pieces for your room is texture. This is especially important when selecting **upholstery** and curtains. **Nubby**, rough textures will create a more casual, homey feel. Smooth and shiny textures such as silk will create a more formal feeling. Though you can mix and match many different **fabrics**, try to stick to those that are complimentary. Avoid **stark contrasts** in texture and instead use contrast to your advantage with

Fig.8-8 Texture

elements such as color. Texture can also extend to the floors and walls. Many types of wallpaper have very distinct textures, and even paint will have varying degrees of gloss[2]. Consider all the elements in a room carefully to create a **cohesive** look and feel.

Form and Line (Fig.8-9)

The next two elements of interior design are often overlooked, they are form and line. In most rooms, your **dominant** line is going to be straight, and is defined by your walls. However you will also find lines in the shape of your furniture as well as the presence of stripes and patterns on your walls or floors. **Vertical lines** create height and tend to add a touch of **formality** to the room[3]. **Horizontal lines** are more restful. A combination of vertical and horizontal lines is usually necessary to create the propersense of balance. **Diagonal lines** will catch your attention, but should be used sparingly as they can create a dizzying effect. **Curved lines** will soften the room and often create a feeling of femininity. Form is the shape created by those lines. The most dominant form in interior decoration is the rectangle, seen in couches and coffee tables[4]. Circular shapes will soften the feel of a rigid room, and triangular shapes create a sense of stability.

Fig.8-9　Form and line

Space (Fig.8-10)

A room can be made smaller or larger with the right use of space. Bright and light colors may make a room appear larger while dark and dull colors may make a room appear smaller. Furniture placed by the wall makes the room appear spacious. By dividing the space in different ways, a room can have an **illusion** of being large or small.

Fig.8-10　Space

New Words

decorate [ˈdekəreɪt] vt. 装饰；点缀；粉刷
texture [ˈtekstʃə(r)] n. 手感，质感，质地
palette [ˈpælət] n. 主要色彩，主色调
disguise [dɪsˈɡaɪz] v. 隐藏，遮盖
compliment [ˈkɒmplɪmənt] vt. 向……道贺；称赞；向……致意
upholstery [ʌpˈhəʊlstəri] n. 家具装饰业；室内装饰品
nubby [ˈnʌbi] adj. 有节的，块状的

fabric [ˈfæbrɪk] n. 织物；布；构造；（建筑物的）结构
cohesive [kəʊˈhiːsɪv] adj. 结成一个整体的
dominant [ˈdɒmɪnənt] adj. 高耸的；突出的
formality [fɔːˈmæləti] n. 礼节；拘谨；正式手续
illusion [ɪˈluːʒn] n. 错觉；幻想；错误观念

Phrases & Expressions

interior design 室内设计
stark contrast 鲜明对比
vertical line 垂直线
horizontal line 水平线
diagonal line 斜线对角线
curved line 曲线

Notes on Text

[1] Each element has its unique characteristics, but once put and mixed together an attractive and creative outcome is produced, thus beautifying the home.
每种元素都有其独特的特点，一旦把它们综合运用，就会创造出别样的吸引力，从而美化房屋。

[2] Many types of wallpaper have very distinct textures, and even paint will have varying degrees of gloss.
许多类型的墙纸都有非常鲜明的纹理，甚至油漆也会有不同程度的光泽。

[3] Vertical lines create height and tend to add a touch of formality to the room.
垂直线可以创造高度，并给房间增添一点仪式感。

[4] The most dominant form in interior decoration is the rectangle, seen in couches and coffee tables.
室内装饰中最主要的形式是长方形，例如沙发和咖啡桌。

Exercises

Ⅰ. Choose the best answers according to the text.

1. According to the passage, color, _____, line, form and space can work together to create the right style of interior design.

    A. adornment   B. painting   C. texture   D. furniture

2. _____ lines will draw you attention but should be used sparingly as they create a dizzying effects.

    A. Vertical   B. Horizontal   C. Diagonal   D. Curved

3. Different kinds of lines shape together to create _____.

    A. form   B. space   C. fabric   D. style

4. Which of the following statement is NOT true in interior design?

    A. The colors you choose should work well together.

    B. Nubby, rough textures will create a more formal feeling.

C. Triangular shapes create a sense of stability.

D. Dark and dull colors may make a room appear smaller.

5. A room can give people an illusion of being big or small _____.

　　A. By dividing the space in different ways.　　B. By creating the space in different ways.

　　C. By combining the space in different ways.　　D. By changing the space in different ways.

Ⅱ. **Fill in the blanks with the following words. Change the forms if necessary.**

| decorate | compliment | texture | cohesive |
| dominant | illusion | interior | vertical |

1. Sloping walls on the bulk of the building create an optical _____.
2. He skillfully fuses these fragments into a _____ whole.
3. Each brick also varies slightly in tone, _____ and size.
4. The late afternoon sun brightened the _____ of the church.
5. For some buildings a _____ section is more informative than a plan.
6. That peak is _____ over all other hills.
7. He addressed her with high _____.
8. Michael was indecisive about how to _____ the room.

Ⅲ. **Translate the following English into Chinese or Chinese into English.**

1. basic element
2. interior design
3. diagonal line
4. vertical line
5. horizontal line
6. 窗帘
7. 鲜明对比
8. 稳固感
9. 圆形结构
10. 曲线

Oral Practices

Conversation One

A: Interior decoration is both an art and a hobby. Nothing is more rewarding or satisfying than building up an elegant interior for our home.

B: That's right. Do you think a home has to be beautiful from the outside as well?

A: One can't have much control over the exterior. But it should fit in with the surroundings and please the passers-by.

B: We spend most of our lifetime at home. And what kind of home do you hope to have?

A: Everyone wants a home aesthetically pleasing and functional. We should aim at combining beauty and comfort with a welcome friendness.

B: I think a well-equipped home makes living more joyful and exciting.

A: It is the touch that makes a house a home. Nobody likes a home, big and well-furnished, but wrecked by a poor color scheme.

B: Buying furniture and decorating home interiors may be a very important investigation for most families. I prefer my home applicable, economical and different.

A: Yes, so do I.

Conversation Two

A: Are you a decorator?

B: Sure, I have done the job for more than twenty years.

A: What do you think about your job?

B: I think decorative works is a part of the building engineering; in the meantime, good decorators are also important to such works.

A: Speaking on good ground, I think so. If there aren't any good decorators, the top quality of decorative works can't be done.

B: Right. High quality decorative works can be done if we have a lot of skillful craftsmen.

A: Sounds reasonable. It also brings endless enjoyment to people. So decorators must try their best to do the job well.

B: Yes, in this way, the quality of the whole decorative works can be guaranteed.

A: Thank you for talking about your job.

B: With pleasure. Thanks for your comprehension and appraisal.

A: It's my pleasure.

Translation Skills—建筑英语翻译之长句的翻译

理解长句大体可以分为两个步骤进行：（1）判断出句子是简单句、并列句，还是主从句；（2）先找出句中的主要成分，即主语和谓语动词，再分清句中的宾语、状语、表语、宾语补足语、定语等。

1. 顺译法

对专业英语而言，一般应尽量采用顺译。顺译有两个好处：一是可以基本保留英语语序，避免漏译，力求在内容和形式两方面贴近原文；二是可以顺应长短句相替、单复句相间的汉语句法修辞原则。

（1）在主谓连接处切断（用"|"表示）。

【例】The main problem in the design of the foundations of a multi-storey building under which the soil settles is | to keep the total settlement of the building within reasonable limits, | but specially to see that the relative settlement from one column to the next is not great.

在土壤沉降处设计多层建筑基础的主要问题，就是要使建筑物的总沉降量保持在合理的限度内，而且特别要注意相邻柱子之间的相对沉降量不能过大。

（2）在并列或转折连接处切断。

【例】The temperature of the water used in mixing the bentonite suspension, and of the suspension when supplied to the borehole, |shall not be lower than 5 °C.

搅拌斑脱岩悬浮物的水温以及灌注到钻孔的悬浮物的温度不应低于5 °C。

（3）在从句处切断。

【例】In the course of designing a structure, you have to take into consideration | what kind of load structure will be subjected to, | where on the structure the load will do what is expected and | whether the load on the structure is applied suddenly or gradually.

在设计结构时必须考虑到：结构将承受什么样的荷载，荷载作用在结构的什么位置，起什么作用，以及这荷载是突然施加的，还是逐渐施加的。

2. 倒译法

翻译时只要能做到顺译，就不一定非倒置不可。在大多数情况下，倒置也只是一种变通手段，并不是唯一可行的办法。

（1）将英语原句全部倒置。

【例】CREC people has built more than 80,000 kilometers long railways, accounting for over two thirds of the total railway line, although there are so many construction groups in China.

虽然中国有很多建筑集团，但是中国中铁先后修建了8万多公里的铁路，占全国铁路总里程的三分之二以上。

（2）将英语原句部分倒置（把句首或首句置于全句之尾）。

【例】It is most important that the specifications should describe every construction item which enters into the contract, the materials to be used and the tests they must meet, methods of constructions in particular situations, the method of measurement of each item and the basis on which payment should be calculated.

对于合同所列的各项施工项目、需要的材料及其检验要求、具体条件下的施工方法、每个施工项目的验收方法以及付款计算的依据等，说明书中都应加以详细说明，这一点十分重要。

3. 拆译法

为了汉语行文方便，有时可将英文原文的某一短语或从句先行单独译出，并利用适当的概括性词语或通过一定的语法手段把它同主语联系在一起，进行重新组织。

【例】The loads a structure is subjected to are divided into dead loads which include the weights of all the parts of the structure, and live loads which are due to the weights of people, movable equipment, etc.

结构物受到的荷载分为静载和活载两种。静载包括该结构物各部分的重量，活载是由人群、可移动设备的重量等所引起的。（增加概括性词"两种"）

Unit Nine Landscape Design
园林设计

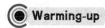 Warming-up

Task Work in pairs. There are ten of the most beautiful landscapes around the world given below. Look at the pictures and then discuss the following topics with your partner.

A. Versailles, France

B. Singapore Botanic Gardens

C. California Fernandez Consol Garden

D. Butchart Gardens

E. Italy Esther Manor

F. Everett Ruth de Rothschild villa garden

G. Dumbarton Oaks

H. Stourhead Landscape Garden

I. Nets Garden

J. Potsdam Sanssouci

Topic 1: What are the similarities and differences between Chinese and foreign landscape gardens?

Topic 2: What kind of elements and principles should be considered during landscape design?

Text A Landscape Design Styles
景观设计风格

An important decision to make before starting your new landscape project is selecting a landscape design style that best matches the look and feel of your home and still reflects your unique personality. This is the best way to create unity between your home and garden. Below are samples of various landscape design styles.

Formal Style (Fig.9-1)

Formal landscape design heavily depends on straight lines and geometrical shapes. The plantings are **orderly** and neatly **pruned** to maintain their formal effect. The Georgian garden would reflect this style and would easily fall under this category.

Fig.9-1　Formal style

Informal Style (Fig.9-2)

Informal landscape design is the exact opposite of the formal style. A more relaxed feel is achieved by using curved lines and irregular shapes. The plantings are massed in a more informal manner creating a **naturalistic** appearance.

Fig.9-2　Informal Style

Tuscan Style (Fig.9-3)

The roots of Tuscan landscaping came from the region of Tuscany located in southern Italy. This style creates an Old World ambiance **reminiscent** of the Italian countryside[1]. The use of stone, old brick, **wrought iron**, heavy wooden beams and **authentic** Tuscan plants are typical of the style.

Fig.9-3　Tuscan style

• 111 •

Mediterranean Style（Fig.9-4）

The Mediterranean landscape design **evokes** visions of **crystal** water and **lush vegetation**, while creating thoughts of relaxation with family and friends and **savory cuisine** made with **aromatic herbs**[2]. The Mediterranean garden style reflects the relaxed Mediterranean culture of southern Europe and combines elegant details with the elements of nature.

Fig.9-4　Mediterranean style

Fig.9-5　English style

English Style（Fig.9-5）

The traditional English garden style has its roots in the English culture. It is most noted for the **array** of **fragrant** flowers and **abundant**, lush plant life, as well as the romantic element if **secluded** sitting areas and **meandering** walkways, covered with creeping vines and majestic shade trees[3].

Tropical Style（Fig.9-6）

Typically **made up of** plants with very large leaves and flowers with intense color, a well designed tropical garden can be very beautiful. Lush **foliage** builds in height towards the back of the garden, creating a dense planting area.

Fig.9-6　Tropical style

Fig.9-7　Asian style

Asian Style（Fig.9-7）

This Asian style of garden design tries to **mimic** nature **on a small scale**. This informality of nature plays a **dominate** role here. Oriental gardens often **incorporate** the art of "feng shui" with its "nine zones" that helps to **instill** a sense of peace and balance to one's life[4].

Contemporary Style (Fig.9-8)

Modern landscape design has been quickly **gaining ground in popularity**. Clean lines, **bold** patterns and new use of materials all play a part of this fresh style. Mass planting in large groups and the use of **abstract** specimens are common practice.

Fig.9-8 Contemporary style

Desert Style (Fig.9-9)

Desert **landscape designs** are popular garden styles for **affluent** communities in the hot and dry areas. The use of native plants that require little or no water flourish is highly recommended. Shade and covered outdoor living rooms are also an integral part of desert landscaping, but the most important thing to remember is to focus on a natural look that provides relief from the intense afternoon sun.

Fig.9-9 Desert style

The above examples are just some of the more popular garden design styles. New and unique styles can be created by combing two or more styles, creating a fusion effect.

New Words

orderly [ˈɔːdəli] *adj.* 有秩序的，整齐的
prune [pruːn] *v.* 修剪
naturalistic [ˌnætʃrəˈlɪstɪk] *adj.* 自然主义的
reminiscent [ˌremɪˈnɪsnt] *adj.* 怀旧的，引起回忆往事的
authentic [ɔːˈθentɪk] *adj.* 真正的
evoke [ɪˈvəʊk] *v.* 唤起，引起
crystal [ˈkrɪstl] *adj.* 透明的，清澈的
lush [lʌʃ] *adj.* 青葱的，繁茂的
vegetation [ˌvedʒəˈteɪʃn] *n.* 植被，植物
savory [ˈseɪvəri] *adj.* 可口的，味美的
cuisine [kwɪˈziːn] *n.* 菜肴
array [əˈreɪ] *n.* 一串，一列
fragrant [ˈfreɪɡrənt] *adj.* 芬芳的，香的

abundant [əˈbʌndənt] *adj.* 丰富的，充足的
secluded [sɪˈkluːdɪd] *adj.* 僻静的，隐蔽的
meandering [mɪˈændə(r)] *adj.* 蜿蜒的，曲折的
tropical [ˈtrɒpɪkl] *adj.* 热带的
foliage [ˈfəʊliɪdʒ] *n.* 树叶，植物
mimic [ˈmɪmɪk] *vt.* 模仿，模拟
dominate [ˈdɒmɪneɪt] *adj.* 支配的，占优势的
incorporate [ɪnˈkɔːpəreɪt] *v.* 使混合或合并
instill [ɪnˈstɪl] *vt.* 逐渐持续地引入，灌输
bold [bəʊld] *adj.* 勇敢的，大胆的
abstract [ˈæbstrækt] *adj.* 抽象的
affluent [ˈæfluənt] *adj.* 丰富的，丰裕的

Phrases & Expressions

formal style 规整式
informal style 非规整式
wrought iron 熟铁，锻铁
aromatic herb 草本香料植物
make up of 构成，组成

on a small scale 小规模地
gain ground 普及，发展
in popularity 流行
landscape design 园林设计

Proper Names

Tuscan Style 托斯卡纳风格
Mediterranean Style 地中海风格
English Style 英伦风格
Tropical Style 热带风情

Asian Style 亚洲风格
Contemporary Style 现代风格
Desert Style 沙漠风情

Notes on Text

[1] The roots of Tuscan landscaping came from the region of Tuscany located in southern Italy. This style creates an Old World ambiance reminiscent of the Italian countryside.

托斯卡纳景观源自意大利南部的托斯卡纳地区。这种风格营造出一种古式的氛围，让人联想到意大利的乡村。

[2] The Mediterranean landscape design evokes visions of crystal water and lush vegetation, while creating thoughts of relaxation with family and friends and savory cuisine made with aromatic herbs.

地中海景观设计营造了流水潺潺和植被繁茂的景象，同时唤起人们与家人和朋友放松，并品尝由香草制成美味佳肴的想法。

[3] It is most noted for the array of fragrant flowers and abundant, lush plant life, as well as the romantic element if secluded sitting areas and meandering walkways, covered with creeping vines and majestic shade trees.

最引人注目的是芬芳的鲜花和郁郁葱葱的植物，以及在僻静的休息区和蜿蜒的走道上匍匐的藤蔓和雄伟的林荫树所体现的浪漫情调。

[4] Oriental gardens often incorporate the art of "feng shui" with its "nine zones" that helps to instill a sense of peace and balance to one's life.

东方园林往往把"风水"和"九宫"艺术融入其中，借此达到内心的平静和生命的平衡。

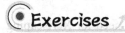

I. Answer the following questions according to the text.

1. What are the main differences between formal style and informal style?
2. What kind of visions does Mediterranean landscape design evoke?
3. In tropical garden, what are the typical plants?
4. What are the main characteristics of contemporary style?
5. How to create a fusion effect in garden design?

II. Match the English words with their Chinese equivalents.

1. reminiscent   A. 支配的，占优势的
2. prune   B. 蜿蜒的，曲折的
3. dominate   C. 唤起，引起
4. affluent   D. 丰富的，丰裕的
5. evoke   E. 真正的
6. authentic   F. 逐渐持续地引入，灌输
7. meandering   G. 修剪
8. instill   H. 怀旧的，引起回忆往事的
9. secluded   I. 使混合或合并
10. incorporate   J. 僻静的，隐蔽的

III. Read the following passage and fill in the blanks with the words given in the box.

| selecting | dynamic | account | involvement | maintenance |
| ongoing | commonly | interrelated | elements | inclinations |

Many landscape designers have an interest and _____ with gardening, personally or professionally. Gardens are _____ and not static after construction and planting are completed, and so in some ways "never done." Involvement with landscape management and direction of _____ garden direction, evolution and care occurring depend on the professional's and client's needs and _____. As with the other _____ landscape disciplines, there can be overlap of services offered under the titles of landscape designer or professional gardener. The followings are the final considerations for _____ landscape design styles.

Consider the design of your home. Is there a strong popular theme that is _____ used and associated with the type of home you have?

Take local natural resources and conditions into _____. Is there a need for Xeriscaping（节水型园艺）, native plants, or a style of landscape design that is more hardscape（硬景观）_____ rather than lawn and a lot of plants?

Future _____. Are you the type that likes to spend a lot of time working in the yard or do you simply enjoy seeing and relaxing in it? While some landscaping styles and themes will work with almost any type of home, some require a lot of work and some require very little at all.

Ⅳ. Translate the following English into Chinese or Chinese into English.

1. intense color
2. Tuscan Style
3. geometrical shape
4. clean line
5. wooden beam
6. 园林设计
7. 规整式
8. 曲线
9. 热带风情
10. 不规则形状

Text B Famous Residential Landscape Design
著名的住宅景观设计

Jesus Galindez Slope and Pau Casals Plaza (Fig.9-10)

Before the **intervention**, the place was a rocky slope ramp trapped within the growing city, **barren** and useless land. The slope is reshaped by using **triangular planes** of different materials that show its strange **topography** to the city. The slope of rock becomes a connecting element: the rock-carved planes of different materials are a lying sloping stair linking the two levels of the neighborhood and collect a large **pedestrian** traffic, which are the alternative to the existing paths around the slope, much longer[1]. The designers take the condition of the place as a high point of the city to create a **balcony** to Bilbao on a raised platform. They removed the old cross between the avenues of Jesus Galindez and Pau Casals and created a wooded area, where the **lime trees** were replanted in existing large size. On the old electrical substation is created a children's play area, integrating the

Fig.9-10　Jesus Galindez Slope and Pau Casals Plaza

platform into an artificial topography of **floor plans** and flexible plants.

Parkview Eclat (Fig.9-11)

Parkview Eclat is a high-end **residential** development with a water-themed landscape to complement the Art-deco style of its architectural style. The outdoor spaces are organized along two major axes. A main axis runs from the main building to the landscape deck, **punctuated** by water fountains and exquisite sculptures. This continues across the **horizontal plane** and even **vertically** up the building's facade, lending coherence to the overall Art-deco look. A second axis categorizes

Fig.9-11   Parkview Eclat

various water-related activities, distinguishing facilities like the water play fountain, children pool, lap pool and Jacuzzi. The linear swimming pool acts as visual **anchor** along this axis, which ends dramatically in the raised Jacuzzi pavilion[2]. The designer smartly combines the modern **exotic** landscape lighting, **art deco landscape style**, fountain and the stone sculpture so that the magnificent of modern landscape design and architecture is produced.

Beijing Times Mansion (Fig.9-12)

Featuring a Chinese modern imperial garden, magnificent and exquisite, Beijing Times Mansion is a beautiful urban landscape. The design concept of Siteline Environment Design is "to integrate Chinese modern imperial garden into the urban life." Under the condition of continuing the city's historical culture and the general urban style, the designers inject new energy to the landscape design. The designers choose the excellent scenery and unique craft and art in Chinese garden to express the garden[3]. Not constrained in the tradition, the overall layout applies an **aberrant** axis to arrange the space. Waterscapes, screenings, borrowed sceneries and opposite sceneries are combined with Chinese Fengshui theory. Both the magnificence of Chinese modern imperial garden and the energy of modern urban life have been strengthened[4]. In the plan of front court, patio and backyard, the winding pathways, the rich plants, the convenient play zone, the waving topography and changing walkways all work together to create a natural and comfortable living space.

Fig.9-12   Beijing Times Mansion

Altamira Ranch (Fig.9-13)

At Altamira Ranch, the architectural and landscape designs work together to create a project that, while impressive in scale, looks and feels as though it emerged from the surrounding environment. This illusion is achieved through the use of building materials that either are or resemble indigenous stone and an almost 100% native California plant **palette**. The planting design in particular connects the built environment not only to the natural but also to the ocean. The drifting masses of native **shrubs** evoke the waves below, paralleling their direction, creating a similar visual rhythm, and reflecting similar colors of greens and blues. Just as waves become shallower and encounter sand bars as they come into shore, the plants become shorter and are interrupted by large areas of sand as they approach the house[5]. At the guest house, the most inland of the structures, succulents that resemble sea **urchins**, star fish and **corals** are used to evoke the feeling of entering a tidal pool or **estuary**[6].

Fig.9-13　Altamira Ranch

New Words

slope [sləʊp] n. 斜坡
intervention [ˌɪntəˈvenʃn] n. 介入，调停，妨碍
barren [ˈbærən] adj. 贫瘠的
topography [təˈpɒɡrəfi] n. 地形学，地形测量学；地貌
pedestrian [pəˈdestrɪən] n. 步行者，行人
balcony [ˈbælkəni] n. 阳台，露台
residential [ˌrezɪˈdenʃl] adj. 住宅的
punctuate [ˈpʌŋktʃueɪt] vt. 加强，强调
vertically [ˈvɜːtɪkli] adv. 垂直地，直立地
anchor [ˈæŋkə(r)] n. 锚，靠山

exotic [ɪɡˈzɒtɪk] adj. 异国的，外来的，吸引人的
aberrant [æˈberənt] adj. 离开正路的，与正确（或真实情况）相悖的
palette [ˈpælət] n. 调色板，颜料
shrub [ʃrʌb] n. 灌木，灌木丛
succulent [ˈsʌkjələnt] n. 肉质植物，多汁植物
urchin [ˈɜːtʃɪn] n. 海胆
coral [ˈkɒrəl] n. 珊瑚
estuary [ˈestʃuəri] n. （江河入海的）河口，河口湾，港湾

Phrases & Expressions

triangular plane 三角形平面
lime tree 椴树，菩提树
floor plan 地面路线图，楼面布置图，楼面图

horizontal plane 水平面
art deco landscape style 装饰艺术风格的景观

Proper Names

Jesus Galindez Slope and Pau Casals Plaza 赫苏斯·加林德斯坡和博·卡萨尔斯广场
Parkview Eclat 景园辉庭

Beijing Times Mansion 北京时代尊府
Altamira Ranch 阿尔塔米拉庄园

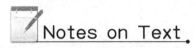

Notes on Text

[1] The slope of rock becomes a connecting element: the rock-carved planes of different materials are a lying sloping stair linking the two levels of the neighborhood and collect a large pedestrian traffic, which are the alternative to the existing paths around the slope, much longer.

布满岩石的斜坡被改造成了一个连接元素：被三角形面板所包裹的岩石上形成了一个阶梯，将两片不同高度的社区连接起来，为斜坡周围的原有道路分散了人流。

[2] The linear swimming pool acts as visual anchor along this axis, which ends dramatically in the raised Jacuzzi pavilion.

线性的游泳池是这条轴线的亮点，泳池的线条在按摩浴缸凉亭那里戛然而止。

[3] Under the condition of continuing the city's historical culture and the general urban style, the designers inject new energy to the landscape design. The designers choose the excellent scenery and unique craft and art in Chinese garden to express the garden.

在延续城市历史文脉和整体城市风格的前提下，设计者给景观设计注入了新的能量，他们采撷中式皇家园林高超的布景、独特的施工工艺等优点来表现园林。

[4] Not constrained in the tradition, the overall layout applies an aberrant axis to arrange the space. Waterscapes, screenings, borrowed sceneries and opposite sceneries are combined with Chinese Fengshui theory. Both the magnificence of Chinese modern imperial garden and the energy of modern urban life have been strengthened.

不拘泥于传统，在造园布局上以错位的景观轴线来布置景观空间，注重水景、障景、借景、对景的运用，并有机地融入风水运势理论，凸显了现代中式皇家园林的大气浩然和现代都市生活的精致与活力。

[5] The drifting masses of native shrubs evoke the waves below, paralleling their direction, creating a similar visual rhythm, and reflecting similar colors of greens and blues. Just as waves become shallower and encounter sand bars as they come into shore, the plants become shorter and are interrupted by large areas of sand as they approach the house.

随风飘动的灌木丛是蓝绿色的，仿佛庄园下方的海浪，灌木的飘动方向和海浪相平行，形成了同样的视觉韵律。正如海浪在靠近沙滩时渐渐变弱，还会撞到沙洲上，庄园周围的植物也渐渐变得低矮，中间还时不时地出现大片的沙地。

[6] At the guest house, the most inland of the structures, succulents that resemble sea urchins, star fish and corals are used to evoke the feeling of entering a tidal pool or estuary.

在最靠近内陆的客房周围种植的多浆植物好像海胆、海星和珊瑚一样，让人想起了海潮池和海口。

Exercises

I. **Choose the best answers according to the text.**

1. Galindez Slope is reshaped by using _____ to show its strange topography to the city.
    A. oval planes   B. quadrangular planes
    C. circular planes   D. triangular planes

2. Parkview Eclat is a high-end residential development with a _____ landscape.
    A. water-themed   B. timber-themed
    C. fountain-themed   D. rock-themed

3. What is the architectural feature of Beijing Times Mansion?
    A. It characterizes an European architecture.
    B. It characterizes a Japan modern architecture.
    C. It characterizes a Chinese modern imperial garden.
    D. It characterizes a Chinese ancient garden.

4. The overall layout of Beijing Times Mansion applies _____ to arrange the space.
    A. a vertical axis   B. an aberrant axis
    C. a horizontal axis   D. a rotational axis

5. The planting design connects the Altamira Ranch to _____.
    A. nature and ocean   B. culture and people
    C. nature and people   D. society and culture

Ⅱ. **Fill in the blanks with the following words.Change the forms if necessary.**

punctuate	estuary	intervention	vertically
barren	residential	aberrant	exotic

1. The government's _____ in this dispute will not help.
2. The place used to be a stretch of _____ land.
3. He made gestures to _____ his speech.
4. The _____ blocks were integrated with the rest of the college.
5. The river opens up suddenly into a broad _____.
6. There are some _____ words in English language.
7. Line the pages for the graph both horizontally and _____.
8. I saw that the insects and spiders were displaying the same kind of _____ behavior.

Ⅲ. **Translate the following English into Chinese or Chinese into English.**

1. an artificial topography　　　6. 变电所
2. landscape deck　　　　　　　7. 视觉韵律
3. imperial garden　　　　　　　8. 地面路线图
4. urban landscape　　　　　　 9. 海潮池
5. an aberrant axis　　　　　　10. 异域景观灯光

Oral Practices

Conversation One

A: What's the contribution of landscape design climate change and to gain more understanding?

B: The fact is that landscape design or landscape architect is very under rated. Or not even known what it is. Majority of people will probably know landscape design as garden design, in the UK and Japan as I know of at least.

A: Yeah, furthermore, landscape design can psychologically change people's behavior and lifestyle by placing right things at right place to give people certain emotion.

B: I always wanted to discuss how much can landscape designers do in achieving sustainable life, or may be more simple and earth-friendly life.

A: Well, I have been educated at university how to design landscapes and the importance of ecology, but not much about what is needed to adapt to climate change or urban life, because we are not ecologists or scientists, but my interest has always been in the climate change.

B: I like to ask if any other sector has seen the potential value in landscape design, and if so what kind of potential?

A: Maybe we can place this topic on the website to see what others think.

B: That's a good idea, I can't wait to start it, Let's go!

Conversation Two

A: Landscape can be designed like architecture. But I am only a graduate student, so I want to discuss and ask about the idea of landscapes with you since you are an environmental manager.

B: It is a space, with people, elements and environment. But it is not a puzzle that can be moved around and fit into places just for the spatial comfort. It is more like reversi (the game).

A: You mean if something goes wrong in such a small space, can it change a whole thing?

B: Yes. So I am looking at more of the positive side because one little thing can change a whole thing as a system.

A: Investigating on landscape is absolutely great I think, just because I had some time to think, I thought there may be some significant connections in all those great designs that I have not noticed.

B: That's right! There is indeed a lot of potential in landscape design. We are now seeing projects coming in with rain gardens, detention ponds, permeable sidewalks and parking areas, improved drainage and landscaping to address the storm water run off. In addition to providing a lovely landscape, these new earth-friendly design ideas help protect our environment.

A: That sounds very interesting! There are so many big ideas and big structures out there now.

B: Absolutely! Good landscape provides better use of our recourses, while also creating more sustainable products, jobs and protecting our environment. It's a win-win situation that makes so much sense in so many ways!

Translation Skills—建筑英语翻译之特殊句型的翻译

1. 被动句型

（1）译成汉语主动句。

①原文中的主语在译文中作主语，将被动语态的谓语译成"加以……""是……的"等。

【例】Distances between elevations *are measured in a horizontal plane*.

高程之间的（斜）距离是用其水平投影来测量的。

②原文中的主语在译文中作宾语，将英语句译成汉语的无主语句，或加译"人们""我们""大家""有人"等词作主语。

【例】*Attempts are also being made to* produce concrete with more strength and durability, and with a lighter weight.

目前仍在尝试生产强度更高、耐久性更好，而且重量更轻的混凝土。

（2）译作汉语被动句。

【例】The cost of a project is *influenced* by the requirements of the design and the specifications.

工程造价会受到设计和技术要求的影响。

（3）把被动句译成汉语的无主句。

【例】In almost every branch of civil engineering and architecture, extensive use is *made of* reinforced concrete for structures and foundations.

几乎在土木工程和建筑的每一个领域，都广泛使用钢筋混凝土结构及基础。

2. 否定句型

英语的否定句多种多样，有一些特殊的否定句，它们所表达的含义、逻辑等都和我们从字面上理解的有很大的差别，值得特别重视。

（1）否定成分的转译。

否定成分的转译是指由意义上的一般否定转为其他否定，反之亦然。常见的句型如下：

①"not... so... as"结构。在谓语否定的句子中，如果带有so... as连接的比较状语从句，或as连接的方式状语，就应该译成"不像……那样……"，而不能直译成"像……那样不……"。

【例】The sun's rays do not warm the water *so much as* they do the land.

太阳光线使水温增温不如它们使陆地增温那样高。

②"not... think（believe）..."结构。表示对某一问题持否定见解的句子，译为"认为……不……""觉得……不是……"。

【例】Ordinarily, we *don't think* concrete as having high tensile.

我们通常不认为混凝土具有很强的拉伸性。

③"not... because..."结构。在汉语中要注意的是，这种结构可以表示两种不同的否定含义，既可以否定谓语，也可以否定原因状语从句。翻译成汉语时，应根据上下文的意思来判断。

【例】This version is *not* placed first *because* it is simple.

这个方案并不能因为简单而放在首位。

这个方案因为太简单，所以不能放在首位。

（2）部分否定。

英语中all, both, every, each 等词与not 搭配使用时，表示部分否定。

【例】*All* these building materials *are* not good products.

这些建筑材料并不都是优质产品。（不能译成"所有这些建筑材料都不是优质产品。"）

类似的结构还有："not... many"（不多），"not... much"（一些），"not... often"（不经常），"not... always"（别总是）等。

（3）意义否定。

【例】The analysis is *too complicated to* complete the computation on time.

分析工作太复杂，难以按时完工。

常见的词组还有：

but for 如果没有

in the dark 一点也不知道

free from 没有，免于

safe from 免于

short of 缺少

far from 远非，一点也不

in vain 无效，徒劳

but that 要不是，若非

make light of 不把……当一回事

fail to 没有

（4）双重否定。

【例】There is *no* material but will deform more or less under the action of force.

在力的作用下，没有一种材料受力不或多或少地发生变形。

常见的搭配还有：

not... until 直到……之后，才能

never... without 每逢……总是

not（none）... the less 并不……就不

not a little 大大地

3．强调句型

（1）强调句型"It is（was）＋被强调部分＋that（which, who）..."几乎可强调任何一个陈述句的主语、宾语或状语。

【例】*It is* these drawbacks *which* need to be eliminated and *which* have led to the search for new methods of construction.

· 124 ·

正因为有这些缺点需要清除，才导致了对施工新方法的研究探求。

【例】*It is* this kind of steel *that* the construction worksite needs most urgently.

建筑工地最急需的正是这种钢。

（2）"It is（was）not until＋时间状语＋that..."是强调时间状语常见的一种句型，可译成"直到……才……"。

【例】*It was not until 1936 that* a great new bridge was built across the Forth at Kincardine.

直到1936年才在金卡丁建成一座横跨海口的新大桥。

（3）在强调句中，被强调的部分不仅可以是一个词或词组，还可能是一个状语从句。

【例】*It is not until the stiff concrete can be placed and vibrated properly to obtain the designed strength in the field* that the high permissible compressive stress in concrete can be utilized.

只有做到在工地正确灌注振捣干硬性混凝土并使之达到设计强度时，才能充分利用混凝土容许压应力。

Unit Ten   Ecology and Architecture
生态与建筑

Warming-up

Task: Work in pairs. In the table there are some common principles to construct ecological buildings. Examples of some famous green buildings are given below. Discuss with your partner and match them with their corresponding principles.

A.

B.

C.

D.

E.

F.

Construction principle	Corresponding picture
Sunlight Efficiency	
Plants Efficiency	
Water Efficiency	
Wind Efficiency	

Text A Green Building
绿色建筑

Green building (also known as green construction or sustainable building) refers to a structure and using process that is environmentally responsible and resource-efficient throughout a building's life-cycle: from siting to design, construction, operation maintenance, **renovation** and **demolition**. This requires close cooperation of the design teams, the architects, the engineers, and the clients at all project stages. The Green Building practice expands and **complements** the classical building design concerns of economy, **utility**, durability and comfort.

High energy costs, environmental concerns and anxiety about the "**sick building syndrome**" associated with the sealed-box structures of the 1970s all helped to **jump-start** the green-architecture movement[1]. Green building brings together a vast array of practices, techniques and skills to reduce and ultimately **eliminate** the impacts of buildings on the environment and human health. It often emphasizes **taking advantage of** renewable resources, e.g., using sunlight through passive solar, active solar, and **photovoltaic equipment**, and using plants and trees through green roofs, rain gardens and reduction of rainwater run-off. Many other techniques are used, such as using low-impact building materials or using packed gravel or **permeable** concrete instead of conventional concrete or asphalt to enhance **replenishment** of ground water[2].

While the practices or technologies employed in green building are constantly **evolving** and may differ from region to region, fundamental principles persist from which the method is derived: Siting and Structure Design Efficiency, Energy Efficiency, Water Efficiency, Materials Efficiency, Indoor Environmental Quality Enhancement, Operations and Maintenance Optimization, and Waste and Toxics Reduction[3]. The essence of green building is an optimization of one or more of these principles. Also, with the proper **synergistic** design, individual green building technologies may work together to produce a greater **cumulative** effect (Fig.10-1).

On the **aesthetic** side of green architecture or sustainable design is the philosophy of designing a building that is **in harmony with** the natural features and resources surrounding the site. There are several key steps in designing sustainable buildings: specify "green" building materials from local sources, reduce loads, **optimize** systems and generate on-site renewable energy[4] (Fig.10-2).

Fig.10-1   Taipei 101   Fig.10-2   US EPA Kansas City Science & Technology Center

The most criticized issue about constructing environmentally friendly buildings is the price. Photo-voltaics, new appliances and modern technologies tend to cost more money. **Proponents** of green architecture argue that the approach has many benefits (Fig.10-3). In the case of a large office, for example, the combination of green design techniques and clever technology can not only reduce energy consumption and environmental impact, but also reduce running costs, create a more pleasant working environment, improve employee's health and productivity, reduce legal **liability**, and **boost** property values and rental returns[5].

Fig.10-3   Green Roof Middle School in France

New Words

renovation [ˌrenəˈveɪʃn] n. 翻新，修复
demolition [ˌdeməˈlɪʃn] n. 毁坏，破坏，拆毁
complement [ˈkɒmplɪment] vt. 补足，补充
utility [juːˈtɪləti] n. 功用，效用
eliminate [ɪˈlɪmɪneɪt] vt. 排除，消除
permeable [ˈpɜːmɪəbl] adj. 可渗透的，具渗透性的
replenishment [rɪˈplenɪʃmənt] n. 补给，补充
evolve [ɪˈvɒlv] vt. 使发展，使进化
synergistic [ˌsɪnəˈdʒɪstɪk] adj. 增效的，协作的，互相作用（促进）的
cumulative [ˈkjuːmjələtɪv] adj. 累积的，渐增的
aesthetic [iːsˈθetɪk] adj. 美的，美学的
optimize [ˈɒptɪmaɪz] vt. 使最优化，使尽可能有效
proponent [prəˈpəʊnənt] n. 支持者，拥护者
liability [ˌlaɪəˈbɪləti] n. 责任，倾向
boost [buːst] vt. 促进，提高

Phrases & Expressions

sick building syndrome 病态建筑综合征，大楼病综合征（指因办公楼空气不好而引起的头痛、眼睛疼痛、疲劳等症状）
jump-start 起动，发动
take advantage of 利用
photovoltaic equipment 光伏设备
in harmony with 与……协调，与……一致

Notes on Text

[1] High energy costs, environmental concerns and anxiety about the "sick building syndrome" associated with the sealed-box structures of the 1970s all helped to jump-start the green-architecture movement.

高昂的能源成本、对环境问题的关注和与20世纪70年代盒状封闭式建筑相关的"病态建筑综合征"所引发的忧患共同推动了绿色建筑运动的兴起。

[2] Many other techniques are used, such as using low-impact building materials or using packed gravel or permeable concrete instead of conventional concrete or asphalt to enhance replenishment of ground water.

许多其他技术也被应用到绿色建筑中，如使用低冲击力的建材或使用填充碎石或透水混凝土取代传统的混凝土或沥青，以提高地下水的补给。

[3] Siting and Structure Design Efficiency, Energy Efficiency, Water Efficiency, Materials Efficiency, Indoor Environmental Quality Enhancement, Operations and Maintenance Optimization, and Waste and Toxics Reduction.

提高选址和结构设计效率，节能，节水，提高材料效率，提高室内环境质量，运营和维护的优化，以及减少废物和有毒物质排放。

[4] There are several key steps in designing sustainable buildings: specify "green" building materials from local sources, reduce loads, optimize systems, and generate on-site renewable energy.

在设计可持续建筑时有几个关键步骤：从当地资源中指定"绿色"建材，减少负载，优化系统，并生成即时可再生能源。

[5] In the case of a large office, for example, the combination of green design techniques and clever technology can not only reduce energy consumption and environmental impact, but also reduce running costs, create a more pleasant working environment, improve employee's health and productivity, reduce legal liability, and boost property values and rental returns.

以一间大办公室为例，绿色设计工艺与精妙技术两者的结合不仅能减少能源消耗和对环境的负面影响，而且能降低运营成本，创造一个更为愉快的工作环境，促进雇员的身体健康，提高生产率，减少法律责任以及提高地产价值和租赁收入。

Exercises

Ⅰ. **Answer the following questions according to the text.**

1. What is the life circle of a building?
2. What are the concerns of Green Building practice in expanding and complementing the classical building design?
3. Why does green building come into popularity?
4. What is the aesthetic principle in green building design?
5. Why do people criticize the construction of environmentally friendly buildings?

Ⅱ. **Match the English words with their Chinese equivalents.**

1. complement   A. 责任，倾向
2. utility   B. 使最优化，使尽可能有效
3. renovation   C. 累积的，渐增的
4. boost   D. 排除，消除
5. liability   E. 翻新，修复
6. cumulative   F. 补给，补充
7. synergistic   G. 美的，美学的
8. optimize   H. 功用，效用
9. aesthetic   I. 促进，提高
10. eliminate   J. 互相作用（促进）的

Ⅲ. Read the following passage and fill in the blanks with the words given in the box.

| adopt | confidence | increased | minimum | credits |
| generally | determine | conservation | created | rating |

As a result of the _____ interest in green building concepts and practices, a number of organizations have developed standards, codes and _____ systems that let government regulators, building professionals and consumers embrace green building with _____. In some cases, codes are written so local governments can _____ them as bylaws (规章, 章程) to reduce the local environmental impact of buildings.

Green building rating systems such as BREEAM (United Kingdom), LEED (United States and Canada), DGNB (Germany) and CASBEE (Japan) help consumers _____ a structure's level of environmental performance. They award _____ for optional building features that support green design in categories such as location and maintenance of building site, _____ of water, energy, and building materials, and occupant comfort and health. The number of credits _____ determines the level of achievement.

Green building codes and standards, such as the International Code Council's draft International Green Construction Code, are sets of rules _____ by standards development organizations that establish _____ requirements for elements of green building such as materials or heating and cooling.

Ⅳ. Translate the following English into Chinese or Chinese into English.

1. environmentally friendly buildings
2. sick building syndrome
3. fundamental principle
4. cumulative effect
5. energy efficiency
6. 可持续建筑
7. 光伏设备
8. 可再生资源
9. 提高室内环境质量
10. 能源消耗

Text B Pushed Slab
推板办公楼

The Pushed Slab is located between two completely different urban **grids**: the **dense** city fabric of blocks and streets in the North and the loose urban fabric in the south with its clear defined and straightforward **infrastructure**. The design is based on the requested office program and the energy requirements. The project combines proven energy **efficiency** technologies with individual office floors and outside spaces, such as **patios**, **balconies** and a garden. The building is highly flexible offering three cores and a central lobby. It can be rented out to one or various tenants without structural changes.

The building is located on a former rail **embankment** of approximately 4,000 m² (Fig.10-4). The volume follows the site restrictions with a slab shaped volume of 150 m long and 21 m wide. An opening in the volume **preserves** the view of a historic building. To create this urban window and to enhance the urban quality of the neighbourhood, the slab is "pushed" until it breaks, then twisted and pushed to the south. This pushing act creates a **distortion** of the floors, offering multiple **terraces** which can be directly accessed from the work areas as well as from the external staircases[1]. The urban window offers a large terrace on the second level. The terrace and the balconies are furnished with trees planted in large pots, offering employees a friendly environment to relax.

Fig.10-4 Location of the building

The building has two faces (Fig.10-5): a calm side in dialogue with the urban fabric of the north side of Paris, and a more **dynamic** side facing south, **rectangular** to the **boulevard**. The building is wrapped in a skin of wood. The windows form a **rhythmic** ribbon, offering optimal sun and light control of the inner spaces. To **contribute to** the **sustainable** development and **taking** the impact of **deforestation into account**, FSC **certified** wood is used[2].

Fig.10-5 Two faces of the building

Pushed slab is an **exemplary** combination of high energy efficiency, economic reality and architectural quality. The added demand of a preserved view line gave the project its exciting shape. It respects the surrounding neighbors and opens up its heart for a collective meaning.

The "Pushed Slab" is the first realization of Paris' first eco-quartier (Fig.10-6). 264 **photovoltaic panels** on the roof will **generate** 90 MW/year, and a grey water circuit will be applied. 22

solar thermal collectors will generate 45% of the energy needed to heat the water. **Sun blinds** are integrated in the south facade and in the cuts. The building is insulated from the outside in order to reduce thermal bridges. The **accumulation** of these proven reliable techniques results in a highly efficient low-energy building which leads to an energy consumption of 46 kWh per m^2/year[3]. Achieving a BBC Effinergie energy label and **complying with** the objectives **set out** in the "Plan Climat de la ville de Paris."

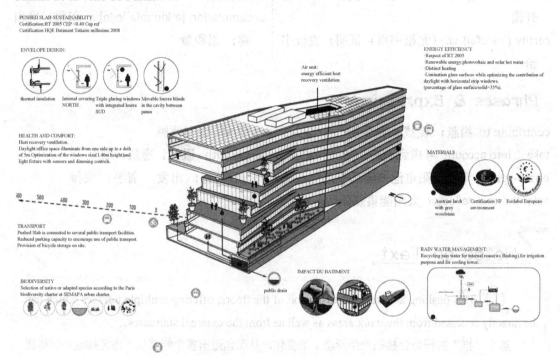

Fig.10-6　Energy saving of the building

New Words

grid [grɪd] n. 格子；（输电线路、天然气管道等的）系统网络
dense [dens] adj. 密集的，稠密的；浓密的，浓厚的
infrastructure [ˈɪnfrəstrʌktʃə(r)] n. 基础设施；基础建设
efficiency [ɪˈfɪʃnsi] n. 效率，效能；实力，能力
patio [ˈpætiəʊ] n. 露台，平台
balcony [ˈbælkəni] n. 阳台；（电影院等的）楼厅，楼座；包厢

embankment [ɪmˈbæŋkmənt] n. 路堤；筑堤
preserve [prɪˈzɜːv] vt. 保护；保持，保存；vi. 保鲜；保持原状；n. 防护用品
distortion [dɪˈstɔːʃn] n. 扭曲，变形；失真，畸变
terrace [ˈterəs] n. 台阶，阶地；阳台；柱廊，门廊；斜坡上房屋间的街巷
dynamic [daɪˈnæmɪk] adj. 动态的；动力的；充满活力的，精力充沛的
rectangular [rekˈtæŋɡjələ(r)] adj. 长方形的，矩形的；成直角的

boulevard [ˈbuːləvɑːd] n. 大马路；林荫大道
rhythmic [ˈrɪðmɪk] adj. 有韵律的，有节奏的；格调优美的
sustainable [səˈsteɪnəbl] adj. 可持续的；可以忍受的；可支撑的
deforestation [ˌdiːˌfɒrɪˈsteɪʃn] n. 采伐森林，森林开伐
certify [ˈsɜːtɪfaɪ] vt. （尤指书面）证明；发证书给……

exemplary [ɪgˈzempləri] adj. 典型的；示范的
panel [ˈpænl] n. 镶板；面；（门、墙等上面的）嵌板；vt. 把……镶入框架内
generate [ˈdʒenəreɪt] vt. 形成，造成；产（后代）；引起
accumulation [əˌkjuːmjəˈleɪʃn] n. 积累；堆积物；累积量

Phrases & Expressions

contribute to 捐献；促成；有助于
take ... into account 考虑到……
photo-voltaic 聚光太阳电池方阵
solar thermal collector 太阳能集热器

sun blinds 太阳薄膜
comply with 服从，遵从；顺应
set out 动身；出发；着手；安排

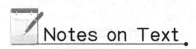

Notes on Text

[1] This pushing act creates a distortion of the floors, offering multiple terraces which can be directly accessed from the work areas as well as from the external staircases.

这个"推"的行为让楼板产生形态上的变化，从而创造出多个能够从工作区抑或外部楼梯都能直达的退台空间。

[2] To contribute to the sustainable development and taking the impact of deforestation into account, FSC certified wood is used.

考虑到森林采伐的影响，并促进可持续性发展，建筑采用的都是通过FSC认证的木材。

[3] The accumulation of these proven reliable techniques results in a highly efficient low-energy building which leads to an energy consumption of 46 kWh per m²/year.

依靠上述所有这些经过验证的可靠科技，能使大楼每年能源消耗控制在每平方米46千瓦时，成为一个高效低能耗的建筑。

Exercises

I. **Choose the best answers according to the text.**

1. What's the main feature of the project?
   A. Flexible office space.　　　　　B. Energy saving.

C. Money saving. D. Both A and B.

2. What does the designer do with the site?

    A. Give up the site. B. Change the site.
    C. Follow the site. D. Not mentioned.

3. What material is the outside surface of the building?

    A. Bricks. B. Wood.
    C. Steel. D. Cement.

4. From the passage, we know that the building is the combination of _____.

    A. high energy efficiency B. architectural quality
    C. economic reality D. all of the above

5. From the passage, we can infer that a good building should _____.

    A. save money B. save energy
    C. respect surroundings and people D. respect the company and people

6. How does the project achieve energy saving?

    A. By using high-technology. B. By using more sunlight.
    C. By using green materials. D. By using less materials.

Ⅱ. Fill in the blanks with the given words. Change the form if necessary.

| balcony | distortion | embankment | generate |
| dynamic | efficiency | dense | sustainable |

1. It was achieved with minimum fuss and maximum _____.
2. Audio signals can be transmitted along cables without _____.
3. Try to buy wood that you know has come from a _____ source.
4. The waves washed over the sea _____ with a loud crashing noise.
5. The company, New England Electric, burns coal to _____ power.
6. There were glass doors leading on to this _____.
7. He mixed business and pleasure in a perfect and _____ way.
8. A _____ column of smoke rose several miles into the air.

Ⅲ. Translate the following English into Chinese or Chinese into English.

1. dense blocks   6. 垂直于
2. energy efficiency technologies   7. 集体意义
3. historic building   8. 光伏板
4. multiple terraces   9. 水循环系统
5. external staircases   10. 建筑质量

Oral Practices

Conversation One

A: There are so many environment problems in the world today. Do you think we can really solve them all or will destroy the world?

B: I hope that world leaders can get together and agree on a plan for action before it's too late.

A: We need to solve the problem of air pollution before we destroy the atmosphere. There's lots of clean, modern technologies, but companies in developed countries say it's expensive. Developing countries put more emphasis on economic development than on environment protection.

B: Everyone is looking at the issue in the short term, rather than the long term. It's the same with the destruction of the rainforests. Countries and companies just want the wood. They're not thinking about the long-term damage to the forests. We should also remember that the forests are an important natural habitat for thousands of species of animal and plant life.

A: In other parts of the world, especially in Africa, there is a problem with desertification. Climate change and over-farming are causing farmland to turn into desert. It means that people can't grow enough food.

B: It also means that people sometimes fight over the farmland that remains. Damaging the environment actually leads to conflict between people.

A: Have you ever thought about joining an organization committed to protecting the environment? You can get involved with projects to improve the environment.

B: I think I'd like to do that. I will take the things I learn here back to my country when finish my studies.

Conversation Two

A: Have you heard that Architect Norman Foster discusses his own work to show how computers can help architects design buildings that are green, beautiful and "basically pollution-free"?

B: Of course! His green agenda is so brilliant and awesome!

A: One of the essential things in green is being low cost. It should reach the people, and have contribution in maintaining them.

B: Also the material used for green architectural forms has to be energy efficient and hence bears low carbon emission!

A: Exactly! I really enjoyed the word "celebratory" other than "sacrificial" "win-lose" etc. in particular in his talk and found it very apt.

B: Yeah, that is what this human solution is all about, to celebrate nature and take care of it to the best of our abilities.

A: Because we are obviously related to it, instead of giving up and running after money and resources like self-interested shortsighted freaks.

B: I can't agree with you any more! Maybe we should share the video of his talk with our classmates.

A: All right. Let's get down to it right now!

Translation Skills——建筑英语翻译之数量的翻译

1. 数量的增加

(1) 非倍数的增加（数量的净增）。

数量的净增指的是纯数量的增加，可以直接译出数字。

【例】Total construction industrial output value *increased 120%* in comparison with 2013.

建筑业生产值比2013年增加了120%。

【例】The living room *is bigger than* the bedroom *by* six square meters.

起居室比卧室大6平方米。

(2) 成倍增加（非净增量的增加）。

非净增量的增加，即倍数的增加。基本句型有以下几种：

①A is N times as large (long, heavy, ...) as B.
②A is N times larger (longer, heavier, ...) than B.
③A is larger (longer, heavier, ...) than B by N times.
④A is N times B.

均可译成：A的大小（长度，重量，……）是B的N倍，或A比B大（长，重，……）N-1倍。

【例】This common high-alumina cement costs roughly *three times as much as portland*.

这种普通的高铝水泥价格大致为硅酸盐水泥的三倍。

【例】Line AB *is five times longer than* Line CD.

AB线比CD线长四倍。

【例】The temperature on the site may *be higher in summer than winter by 40 times*.

工地的夏季气温可能是冬季气温的40倍。

【例】Aluminum has nearly *5 times the thermal conductivity of cast iron*.

铝的导热率是铸铁的五倍。

(3) increase 类表示增加的动词 (to, by) + *N* times/ *N*-fold/ by a factor of *N*。

此句型含义也是"增加到原来的N倍"或"增加了N-1倍"。

【例】Such construction procedure can increase productivity *over threefold*.
这种施工工序可使生产率提高3倍以上。（提高了2倍多）
在这个句型中，increase类动词也经常以名词的形式出现。

【例】The composites showed *a two-fold increase* in modulus of elasticity.
这些复合材料的弹性模量比单纯基体材料增大了一倍。

有些单词可直接表示倍数关系，如double（增加一倍、翻一番），treble（增加两倍或增加到三倍），quadruple（增加三倍、翻两番）等，可以不与具体倍数连用。

【例】This country has *trebled* her annual output of steel in the past ten years.
这个国家在过去十年中钢产量增加两倍。

可见，以上所列句型在译成汉语时，都要将句中的倍数"减一"，但是，如果倍数是一个相当大的近似值，差一倍没有多大意义时，往往可以照译，不必"减一"。

【例】The sun is *330,000 times as large as* the earth.
太阳是地球的330 000倍。/ 太阳比地球大330 000倍。

（4）数量增加的特殊表达法。

①英语中，"比……大一半"有其特殊的表达方式：
half as much（many, large, fast 等）again as
或 half again as much（many, large, fast 等）as
上面两个句型表示"大一半"或"是……的一倍半"。

【例】The resistance of aluminium is approximately *half again as great as* that of copper for the same dimensions.
尺寸相同时，铝的电阻为铜的一倍半。

②在表示"是……的两倍"时，也有多种表达方式：
as much（many, large, fast 等）again as
again as much as double
以上句型表示"大一倍"或"是……的两倍"

【例】Wheel A turns *as fast again as* Wheel B.
A轮转动速度比B轮快一倍。

2．数量的减少

（1）数量的净减。
纯粹数的减少指具体数量的减少、百分比的减少。数字照译，通常译成"减少了N"。

【例】The temperature *fell to* 26 degrees below zero.
温度降到零下26摄氏度。

【例】The pressure will *be reduced to* one-fourth of its original value.
压力将减少到原来数值的1/4。

· 138 ·

（2）成倍减少。

语句中表示成倍减少含义时，通常包含以下句子成分：

reduce by N times/reduce N times as much（many...）as

reduce N times/reduce by a factor of N

reduce to N times/reduce N-fold

N-fold reduction/N times less than

均可译成"减少了（N–1）/N"或"减少到原来的1/N"。

例如，reduce 3 times 可译为"减少了2/3"或"减少到原来的1/3"。

【例】The construction cost *has reduced four times*.
建筑成本减少了3/4。（即减少到原来的1/4）

【例】The advantage of the present scheme lies *in a fivefold reduction* in manpower.
这一方案的优点在于节约人工4/5。（即节约到原来的1/5）

（3）decrease 类动词（to, by）+N times（N-fold）。

该句型意义是："减少了1–1/N"或"减少到原来的1/N"。

【例】The prices of cement have been *reduced four times* as compared with 1950.
水泥价格与1950年相比降低了75%。

【例】There has been *a five-fold decrease* in the loss of metal this year.
我们今年的金属损耗率减少了4/5。

【例】The output of strip steel has been *reduced by 2.5 times*.
钢带产量已降低到原来的2/5。

（4）数量减少的特殊用法。

在表示少一半时，英语往往习惯用 "half as... as" 或 "twice less than" 来表达，一般不用two times.

【例】The new compressor is *half as heavy as* the old one.
这部新压缩机比旧的轻一半。

专业英语中，减少一半还有以下表达法：

to decrease one-half, to reduce by one-half, to cut in half, to shorten twice, one-half less 等。

3. 不确定数量

英语中常用修饰不确定数量的词有：circa（拉丁语），about, around, some, nearly, roughly, approximately, or so, more or less, in the vicinity of, in the neighborhood of, a matter of, of the order of 等。这些词可译成"大约……""接近……""……上下""……左右"等。

【例】a weight *around* 12 tons   12吨左右的重量
300 *or so*   大约300米

Appendix I：课后习题答案

◎第一单元

Warming-up

Text A Exercises

Text B Exercises

◎第二单元

Warming-up

Text A Exercises

Text B Exercises

◎第三单元

Warming-up

Text A Exercises

Text B Exercises

◎第四单元

Warming-up

Text A Exercises

Text B Exercises

◎第五单元

Warming-up

Text A Exercises

Text B Exercises

◎第六单元

Warming-up

Text A Exercises

Text B Exercises

◎第七单元

Warming-up

Text A Exercises

Text B Exercises

◎第八单元

Warming-up

Text A Exercises

Text B Exercises

◎第九单元

Warming-up

Text A Exercises

Text B Exercises

◎第十单元

Warming-up

Text A Exercises

Text B Exercises

Appendix Ⅱ：课文参考译文

第一单元 世界建筑史

·世界建筑史简介·

早期的建筑

古埃及从开始使用砖石到最后建成宏伟的金字塔和巨大的狮身人面像，已成为历史上许多有影响力建筑物的故乡。尼罗河流域在至少一万年间，一直都是艺术、建筑、设计这些灿烂文明的摇篮，这方面的创新均源于一个简单的原因：缺少木料。

埃及人是最早利用经久耐用的砖结构的群体之一，他们的建筑古迹历经数千年，甚至从现代来看也是古代建筑的典范。正因为砖石的经久耐用，史上一些最著名的建筑均源自埃及的尼罗河流域。

希腊、罗马时期的建筑

随着西方社会的繁荣和发展，希腊人的建筑设计呈现出新的活力。一个强大的文明在欧洲异军突起，希腊建筑师凭借他们在建筑法则、设计和美学方面的才华，在建筑史上留下辉煌的一页。希腊人还创造了关于建筑美学的最初标准以及理想的比例，这些典范都被随后的其他国家所借鉴。

希腊两种建筑柱式风格差不多是同时发展起来的。希腊本土和西方殖民地以多立克柱式为主。爱奥尼柱式起源于小亚细亚的岛屿和沿海城市。

罗马是追随希腊步伐最主要的国家，继希腊风格后创造了世界史上最有名的一些建筑。以一些伟大的工程、道路、运河、桥梁和渡槽为标志，罗马发展成为强盛的、组织有序的帝国。两项罗马发明——圆屋顶和交叉拱，使建筑风格更加灵活多变。罗马人还建造了纪念碑、凯旋门和斗兽场以及体育场。

文艺复兴时期的建筑

文艺复兴在公元1400年左右从意大利开始，它带动了古希腊和罗马建筑的原则与风格的复兴，在之后的150年间蔓延到欧洲其他国家。文艺复兴期间，建筑几乎取得了最伟大的飞跃。和谐的外形、精确的数学比例与优美的格调结合在一起，让我们感受到许多创新之处，比如彩色玻璃窗、哥特式教堂、高耸的尖顶，当然还有八角形的穹顶。

巴洛克和洛可可建筑

文艺复兴时期后，那个时代的建筑师开始厌倦在过去200多年来一直使用的对称和相同的旧形式。巴洛克最鲜明的风格是采用了椭圆形的外形，并出现在许多教堂上，成为当时最主要的特色。

洛可可是法国巴洛克的最后阶段，是一种意于迎合巴黎人的轻松的用于装饰的风格。洛可可风格的一大主要特色是用于演奏音乐的房间设计。

工业时代的建筑

1760年始于英国的工业革命让很多新的建筑材料涌现出来，比如铸铁、钢和玻璃。18世纪末的设计师和赞助商倾向于最初的希腊和罗马风格。在19世纪，英国建筑师约瑟夫·帕克斯顿爵士在伦敦创建了一个巨大的展览厅——水晶宫（1850—1851），预示着工业建筑以及铸铁和钢材的广泛使用。

现代建筑

20世纪初，一些设计师拒绝借鉴他人的风格。西班牙建筑师安东尼·高迪是最具代表性的原创设计师。

大约在1965年到1980年，建筑师和评论家开始拥护后现代主义。虽然后现代主义不是一项基于一套独特的原则有凝聚力的运动，但是通常后现代主义者更看中个性、亲密性、复杂性以及偶尔的幽默。

到了20世纪80年代初，后现代主义已经成为美国建筑的主流趋势，在欧洲也成为一个重要现象。

·世界"新七大奇迹"·

受古代世界七大奇迹的启发，某民间组织决定从候选名单中选出七个新的世界奇观。2007年7月7日新的世界七大奇迹在葡萄牙里斯本被提名。下面是由世界各地人民从21个候选景点中选出的世界新七大奇迹。

长城，中国（公元前220年和公元1368年——公元1644年）

中国的万里长城是中国古代为抵御蒙古部落入侵而建造的，它将已有的单个要塞连成一体，从而形成一个完整的防御体系。长城是世界上最大的人造工程，也是宣称唯一可以从外太空看到的地球景观。为了修建巨大的长城，成千上万人为此献出了生命。

佩特拉古城，约旦（公元前9年——公元40年）

在阿拉伯沙漠的边缘，佩特拉古城曾是国王阿莱塔斯四世（公元前9年——公元40年）当政时的纳巴泰王朝的首都。纳巴泰人是水设施技术方面的杰出大师，他们为佩特拉修建了伟大的水渠和蓄水池等供水系统。佩特拉城中有一个大竞技场，以希腊罗马的建筑为蓝本而建，最多可容纳4 000个观众。今天，在埃尔代尔的修道院，佩特拉陵寝与42米高的希腊寺庙面对面矗立，这已成为中东文化遗产的重要代表。

古罗马斗兽场，意大利（公元70年——公元82年）

位于罗马中心的壮观的斗兽场是古罗马当时为取悦凯旋的将士和赞美伟大的古罗马帝国而建造的。斗兽场的建筑设计并不落后于现代的美学观点，而事实上，大约2 000年后的今天，每一个现代化的大型体育场都或多或少地烙上了一些古罗马斗兽场的设计风格。如今，通过电影和历史书籍等媒介，我们能更深切地感受到当时在这里发生的人与兽之间的残酷格斗和搏杀，而这一切，只是为了给作壁上观的观众带来一些原始而又野蛮的快感。

马丘比丘，秘鲁（公元1460年——1470年）

公元15世纪，印加王朝的国王帕查库蒂在马丘比丘山脉（古老山脉）上修建了一座云中之城。这个非凡的居住地上延至安第斯山脉，下潜至亚马孙热带丛林，并位于乌鲁班巴河谷之上。历史上由于"天花"爆发，该城市遭到印加王朝遗弃。当西班牙人打败了印加帝国后，这座古城便"消失"了长达三个多世纪，直至1911年，海勒姆·宾厄姆发现了该城遗址。

奇琴伊察金字塔，墨西哥（公元800年前）

位于墨西哥尤卡塔半岛上的奇琴伊察是玛雅古国最著名的城邦遗址，它曾是古玛雅文明的政治和经济中心。城内至今仍可见的古迹主要有库库尔坎金字塔、查尔穆尔神殿、千柱林和囚犯竞技场。这些建筑在空间和造型组合上均充分体现出了玛雅人杰出灵动的建筑意识。而金字塔本身，则是众多玛雅寺庙中最后也是最杰出的代表作。

泰姬陵，印度（约公元1631年——1653年）

泰姬陵是莫卧儿王朝第五代皇帝沙贾汗为纪念其已故爱妻而下令修建的陵墓。陵墓由白色大理石砌成，坐落于有高大围墙的花园中，非常宏伟壮观。泰姬陵被公认为是穆斯林建筑艺术在印度最杰出、最完美的代表。人们传说，后来沙贾汗皇帝被囚禁起来，每天只能透过小小的囚窗来观望美丽的泰姬陵。

基督像，巴西（公元1931年）

这座高达38米左右的基督雕像坐落在里约热内卢的科尔科瓦杜山上，俯瞰着美丽的里约热内卢。它是由一个叫保罗·兰多斯基的法国雕刻家和一个叫海托·达·席尔瓦·科斯卡的巴西人分别设计和创造的，是世界上最著名的古迹之一。这座巨型雕像的制作共花费了五年时间，于1931年10月12日完成了落成典礼。巨大的基督像张开双臂，欢迎来参观的游人。如今，基督像已成为美丽的里约热内卢和巴西人热情的象征。

第二单元　中国古代建筑

·中国古代建筑简介·

中国作为一个文明古国位于亚洲大陆的东部，陆地面积约960万平方公里，人口接近世界总人口的1/5，拥有56个民族和5 000多年的历史。中国悠久的历史创造了灿烂的古代文明，而独特杰出的中国传统建筑便是其重要的组成部分。

建筑设计理念

所有中国古代建筑都是根据严格的设计规范来建造的，其一直追随道教和其他各学派的传统理念来设计。第一个建造理念是建筑物长而低，而不是短而高——看起来好像是要把人围起来一样；屋顶是由木梁支撑而不是由墙体支撑；屋顶看起来像漂浮在地面上。第二个理念是对称：建筑物两边的结构相同且对称，这恰好是遵循了道教所推崇的平衡原则。即使在公元前1500年前的商朝，建筑也都是相当对称的，都有着上翘的瓦屋面和长排的立柱。

建筑特点

中国古代建筑总体上说以木结构为主，有着独特迷人的外表，这与主要以砖石结构的世界其他国家建筑不同。木质的柱子，横梁和托梁构成房屋的框架。墙壁起分隔房间的作用而不是支撑作用，这是中国古代建筑的独特之处。正如中国民间俗语所说"墙倒屋不塌"。因为木料耐久性差，需要采取专业的防腐措施，后来就发展成中国自己的建筑彩绘艺术。彩色的琉璃瓦屋顶，精湛设计的窗户，以及带有漂亮花朵图案贴花的木柱子都反映了古代工匠们高超的技艺和丰富的想象力。

建筑与文化

建筑与文化是紧密相连不可分割的，在某种意义上可以说建筑是文化的载体。中国古代建筑风格丰富多变，例如庙宇、宫殿、祭坛、楼阁、官邸和民宅。这些不同类别的建筑都体现了中国古代的思想——天人合一的理念。

中国古代建筑文化的两个典型代表是风水和牌坊。

风水：中国传统理念尤其强调建筑要以《易经》文化为基础，强调人与自然的和谐统一。

牌坊：也叫作牌楼，是中国独一无二的封建社会文化的产物，是封建社会为表彰功勋和忠孝节义所建立的建筑物。

今天，以其传统风格为基础的中国建筑吸收了国外建筑文化，并遵循我们时代的要求，采用日益更新的建筑技术不断开拓进取。

·中国建筑史·

中国建筑就像中国文明一样古老，可谓自成一体。考虑到建筑所反映的社会、经济、政治和技术发展历程，可将中国建筑的发展历程分为以下六个主要阶段。

第一阶段：公元220年以前

从原始社会早期，至公元220年汉朝衰败，这一阶段标志着中国建筑从起步发展到早期成熟阶段。天然洞穴是原始人类解决居住问题的自然途径。

公元前206年至公元220年，中国已经进入封建主义完全成熟的阶段。汉朝皇帝的宫殿宏伟壮丽，有"千门万窗"的说法。为了赢得天神的青睐，皇家园林中修建了很多高塔。与西亚贸易的接触，给中国带来西方的文化因素，同样在中国建筑物中有所反映，如拱门，建筑装饰中出现的人形。佛教也在此时传入中国，所有这些来自西方的新元素，都极大地丰富了中国的建筑。

第二阶段：佛教的影响（公元3世纪至6世纪）

本阶段是包括建筑在内的中国文化吸收印度和中亚文化最多、受其影响最大的时期。佛教寺庙和宝塔几乎遍及各地。

这一时期的主要遗存建筑物，包括一些石窟或洞窟以及凤毛麟角的砖石宝塔。其中一些石窟，如云冈石窟、麦积山石窟以及天龙山石窟，建筑处理都达到了较高的水准。中国保存至今最古老的佛塔就出自这一时期。河南省嵩山嵩岳寺塔，建于公元520年，高约40米，雕饰有一些印度图案，标志着一种新的建筑形式在中国的出现。

第三阶段：盛唐时期（公元7世纪至9世纪）

到公元6世纪末，中国再次统一，其绘画、雕刻、音乐、舞蹈均因受西方影响而得以大大丰富。在建筑领域，不管是技术或是艺术方面，均进入了前所未有的全盛时期。中国现存最早的木结构建筑典范就出自唐代，从中可见木匠工艺的精湛程度。此前佛塔多用木造，此时首次以砖石结构取而代之，形式上不断推陈出新，出现了"楼阁式"佛塔和"重檐"佛塔。

第四阶段：大型建筑群标准化和精细化（公元10世纪至14世纪）

这一时期的主要发展在于建筑规模扩大以及手法日趋华丽精细。为节约材料和劳力，并且加快大规模建筑物的设计速度和建造进程，出现了对典型化和标准化的要求。《营造法式》的编撰出版因而成为中国建筑发展史上的一个重要事件。

现存的这一时期的宏伟建筑，数目较大。除寥若晨星的几座桥梁之外，全部为木结构的寺庙和砖石结构的佛塔等宗教建筑。在佛塔平面布局上，形成了鲜明的十字交叉轴线布局。这一时期的佛塔形式多样。布局方面，除了方形，八边形成为此时和之后最受欢迎的平面布局。

第五阶段：北京——欧式影响的时代（公元1368年至1949年）

明清两代留下了不胜枚举的建筑，在北京城两度扩建中得以保存。北京城及其宫殿苑囿，是这一阶段最重要的历史遗迹，成为最宏伟壮丽的建筑作品之一。

19世纪期间，资本主义和帝国主义的西学东渐，带来了西洋建筑理念。一些中国沿海城市，在帝国主义者得到特许的"租界"中，甚至复制了不少19世纪欧洲的城市建筑。

20世纪上半叶的中国建筑如实地反映了中国半殖民地半封建社会鼎盛时期的风貌。在田间乡下和小城小镇，传统中国建筑仍然是人们的选择。但在较大的城市，"洋房"甚为时髦。

第六阶段：人民的建筑时代（1949年后）

1949年，中国人民建立了中华人民共和国，从此开启了中国建筑的新篇章。

第三单元 建筑设计

· 建筑设计简介 ·

建筑设计师往往在建筑地点、建筑类型及建筑造价三者决定之后进行设计。因此,建筑设计是对于环境、用途和经济上的条件和要求加以运筹调整和具体化的过程。这种过程不但有其实用价值,而且有文化价值,因为设计师任何一种社会活动所创造的空间布置,都将不可避免地影响到人们在其中的活动方式。

环境设计

自然环境有其有利和不利因素,设计师在设计过程中需取长补短。他们为了让房子适宜人居住且舒服,就要考虑寒暑、日光、空气、湿度等的影响并预见潜在危害,如火灾、地震、洪水、疾病等因素。

•朝向

建筑物轴线的朝向以及内部空间布置是控制阳光、风雨影响的一种方法。周围环境(树木、地形、水域及其他建筑物)也和朝向密切相关。

•建筑形式

适当的设计可以制约和减轻自然条件的影响。如挑檐遮阳防雨、屋顶排泄或保存雨雪、墙可隔热保温、窗可采光通风等。

•色彩

色彩有控制吸收和反射日光程度的实用功能,同时,也有其美学上的表现作用。

•材料和技术

选择材料时应同时考虑其本身防御自然条件的能力以及对人的利害关系。设计中需权衡其优缺点。

•室内控制

室内空间的大小、形状、连接方法,以及地面、墙面、天花板、家具等所用材料的颜色、质地均可以控制室内的温度、光线、音响等。现在采暖、隔热、调节空气、照明和音响效果等措施已成为建筑设计中的重要内容。

功能设计

环境设计是要满足人们的感觉(视觉、触觉、听觉)和反应上的舒适,而使用功能设计则为了人们行动和休息便利。

•分区

分区应根据不同的功能要求将空间分区。

•活动空间

在不同的功能分区之间和室内外之间建立适当的行动路线。路线应当简捷、目标明显、设备方便。

•设施

应对人体的尺度、活动方式和体力加以研究分析，以决定层高、门窗、踏步等的尺寸，尽量采用便于人们活动的设施。

经济核算设计

建筑的费用主要用于土地、材料和人工。土地有限时，建筑只能向高度发展。建筑材料的选择受到造价的限制时，建筑设计的各个方面都会受到影响。人工的昂贵要求构造的简化和标准化。经济核算设计中不仅要对三者分别核算而且应综合考虑其比例关系以取得最佳方案。

·给现代建筑重新定义的建筑设计师·

罗伯特·文丘里、弗兰克·杰里、安藤忠雄和扎哈·哈迪德被广泛认为是当今世界顶尖的建筑设计师。他们都以不同的方式为现代建筑创造了重要的范例。你可以在全世界的现代建筑设计中看到他们那充满活力和想象力的建筑。

罗伯特·文丘里

自20世纪60年代以来，美国建筑师罗伯特·文丘里的著作和建筑设计为我们重新定义现代建筑找到了一条途径，位于宾夕法尼亚州费城的凡娜·文丘里公寓就是他早期重要的作品之一。

在他的建筑和著作中，罗伯特·文丘里呼吁这样一种现代建筑，其既能表现历史，又能包含现代通俗的文化。

他的建筑还包括英国伦敦的国家艺术走廊扩建部分。文丘里按照非常现代的模式来设计这座大型建筑，这座博物馆位于伦敦特拉法尔广场，是19世纪的主要建筑之一。文丘里还给许多美国大学做过设计，包括哈佛大学和宾夕法尼亚大学。

弗兰克·杰里

弗兰克·杰里认为建筑是一门艺术。他曾经说，在某些方面，他受到的艺术家和雕刻家的影响要比受到建筑师的影响还要大，这也就是为什么他的建筑通常看起来就像是由几何图形形成的充满活力的雕刻品。

他设计的"舞蹈房"位于捷克首都布拉格。这是一座非常有趣的建筑，建成于1996年，它看起来就像是两位翩翩起舞的少女。杰里最著名的建筑是位于西班牙毕尔巴鄂的古根海姆博物馆。这座博物馆于1997年建成，它的绝大部分弯曲处都覆盖着钛金属，看起来就像是河岸跳动的金属波浪。

安藤忠雄

安藤忠雄是一名日本建筑师。1993年，他因Rokko住房工程而获得了日本文化创意奖。安藤忠雄因把未成型的钢筋混凝土作为建筑材料而闻名，而且，他还因以尽可能地与自然环境相融合的简单设计来建造住房而闻名。

2002年，他的位于得克萨斯州福特沃斯堡的现代艺术馆建成开放，这座建筑非常令人惊奇，使你不得不提醒自己是来观看艺术的。这座混凝土和玻璃博物馆就建在水边，所以，它看起来就像在水中浮动一样。

扎哈·哈迪德

出生在伊拉克的扎哈·哈迪德在英国完成了她的建筑学学业，现在定居在那里，2004年哈迪德成为第一位获得普立兹克建筑大奖的女性。

哈迪德因设计和完成在纸上都几乎不可能实现的建筑而闻名于世。她说，建筑自然是为了掩蔽，但同时也能给人们带来愉悦。哈迪德的著名建筑设计包括法国拉斯堡停车场和终点站，俄亥俄州辛辛那提的罗森塔尔当代艺术中心等。

第四单元　建筑材料

·建筑材料简介·

建筑材料是指用于建筑用途的任何材料。许多天然存在的物质，如黏土、岩石、砂、木材，甚至枝叶，都被用于建造建筑物。除天然材料外，很多人造产品也应用在建筑施工中。建筑材料的历史发展趋势从天然走向人工合成，从可降解到抗腐蚀，从本土使用到全球运输，从修复使用到一次性材料，并且选材的防火等级也不断提高。这些发展趋势往往会增加建材从初始到长期的经济、生态、能源和社会成本。

砌体

砌体包括天然材料，如石头，或者制成品，如砖和混凝土块。砌体从古代就开始被使用：泥砖被用在巴比伦城市的非宗教建筑物中，石头被用在尼罗河流域的大寺庙中。埃及的大金字塔是最壮观的砌体建筑。

木材

木材是最早的建筑材料之一，是为数不多的具有良好的抗拉特性的天然材料之一。世界各地可以发现成百上千个不同品种的木材，每一个品种都表现出不同的物理特征。只有少数的种类在建筑施工中作为框架构件在结构中使用。

钢筋

钢筋是一种重要的结构材料。它能通过碾压被做成不同的结构形式，比如I形梁、板和薄板；它还可以被浇铸成复杂的型式。在钢材中加入合金元素会产生高强度的钢筋。这些钢筋被用在结构构件尺寸很重要的结构中，如摩天大楼的柱子中。

铝材

当轻质、高强度和抗腐蚀成为所有的重要因素时，铝就是一种特别有用的建筑材料。因为纯铝非常软且具韧性，所以必须要添加一些合金元素，如镁、硅、锌和铜等，以给它提供结构使用所需要的强度。除了用于建筑物和预制装配式房屋中的框架构件以外，铝更广泛地用于窗框和幕墙建筑结构的外表面。

混凝土

混凝土是水、砂、砾石以及波特兰水泥的混合物。混凝土在拌和、浇筑后,逐渐硬化成一个像石头一样的、高强度的密实块体。混凝土基本上是一种抗压材料,但是抗拉强度是微不足道的。

钢筋混凝土

混凝土的压力很强,拉力却很弱。钢有很强的拉力。所以,混凝土用钢筋加固后,既有很强的压力,又有很强的拉力,这种材料被称为钢筋混凝土。在建筑施工中,钢筋混凝土可被浇铸成各种形状,例如梁、柱、板和拱,因此它易适用于建筑的特殊结构。

塑料

塑料因为它的多样化、强度、耐久性和轻质性,很快就成了重要的建筑材料。塑料是一种合成材料或树脂,它能被浇铸成任何期望的形状,它还使用有机物质作为胶粘剂。

·小蚂蚁皮影剧场·

小蚂蚁皮影剧场是位于北京大栅栏西河沿大街的一个公益设计项目。剧场的使用者是小蚂蚁皮影剧团,一个由生长激素分泌不足患者组织成的民间皮影演出和制作团体。之所以得名"小蚂蚁",是因为这些被称为"袖珍人"的患者只有儿童的身高,而且其相貌也像儿童。这个皮影剧团是这个病患群体自强不息的模范。然而,长期以来却没有合适的固定剧场,因此落户大栅栏这一传统街区,既是找到了他们理想的家,又是找到了一个让社会更多关注"袖珍人"的窗口。

皮影作为一个传统的手工艺,用复杂而细致的雕刻工艺制作戏剧中迷人的人物。室内设计利用了皮影戏这一特点,并将细节雕刻的理念应用到室内设计中。房间的主材选用金黄色的麦秸板,表面有着自然的纹理,同时,也微微散发着麦秸的芳香。这种图案被设计在四道悬梁上,形成了观众区的纵深感。悬梁在侧墙结合立板落地,构成一整面货架。而另一面墙的上半部则设计为通长的镜面,把狭窄空间放大了一倍。舞台被设计为可旋转式,这样便可以在非放映时间打开去往后面工作间的通道。工作间只有两米宽,因此设计了一张曲线轮廓的桌子,满足了人员在两侧入座,同时师傅还可以置身桌面的中心来指导大家的工作。

这个公益项目得到了许多社会资源的支持和奉献。照明设计是由2008年奥运广场的设计团队义务设计的,他们还联系来免费灯具。店面是一块大玻璃的简单构图。这块玻璃是和用于苹果店里面一样的超白高透玻璃。世界上最高的透明度会使内部的活动能最大限度地被过往的路人透视。类似这样的公益支持几乎涉及这个项目的每一种材料。当然,设计更是无私的奉献。

这个设计力图用当下的设计理念,为一个特殊的残疾人群打造新的社会舞台。小蚂蚁皮影剧场建成后得到了使用者和观众的热烈喜爱,使大众更多地了解到可爱的"袖珍人"。当建筑师号召大家为这个公益项目出力得到热烈反响后，又更证明了公益设计作为一种行为，释放出社会公平的正能量。

第五单元　建筑结构

·建筑结构简介·

建筑物中承受重量和荷载的部分称为结构部分，而不承载建筑物重量的部分如窗户等为非结构部分。建筑结构按照不同的材料可以分为很多种形式，如钢筋混凝土结构、砌体结构和钢结构。

钢筋混凝土结构

在钢筋混凝土结构中，利用钢筋被裹在混凝土结构中抵抗较大拉应力来使混凝土良好的抗压强度得以充分发挥。

钢筋混凝土结构具有自重大、刚度高、耐久性好等特点。由于建筑物承受荷载的方式不同，建筑结构可以分为不同的类型，如框架结构、剪力墙结构、框架—剪力墙结构和筒体结构。

•框架结构

由梁、柱构件通过刚性节点连接而成的超过横截面尺寸的承重结构称为框架结构。

•剪力墙结构

混凝土竖向连续墙可以在建筑上起到分割作用，或在结构上起到承受荷载和抗风作用。在剪力墙结构中，剪力墙同时用来抵抗建筑物承受的侧向荷载。

•框架-剪力墙结构

当剪力墙结构与框架结构结合，剪力墙就接近于弯曲变形，而框架则趋于剪切变形，他们都是被大梁和板的水平刚度所约束而表现出共同的挠曲形态。剪力墙和框架水平向的相互作用，特别是在结构顶端，将产生更高的刚度和强度。

•筒体结构

只有将所有竖向构件相互连在一起，高层建筑所有结构类型中用于抵抗风荷载的强度和刚度才能达到最优值。这样整个建筑就像一个伸出地面的中空筒或刚性盒子，这样的建筑结构就是筒体结构。

砌体结构

最早使用砌体结构的国家可追溯到2000年前的中国。由于较好的隔热性能并易于施工，砌体目前被作为主要的住宅建筑材料。但是由于砌体材料如砖或砌块是一种脆性材料，其抗压能力强，但拉伸能力弱，变形性或延性都较差。

钢结构

钢结构是一种应用广泛的建筑结构形式，其中钢材发挥着主导作用。钢材提供了比混凝土更高的抗压和抗拉性能，并且可以使建筑物变得更轻。钢结构可使用空间桁架，因此，它们能实现比混凝土结构更大的跨度。

·牛背山志愿者之家·

牛背山志愿者之家是一个公益设计，为一群年轻的志愿者们在大山里盖一座房子，地点在四川省甘孜藏族自治州泸定县蒲麦地村。牛背山，被誉为中国最美观云海之地，每年有很多驴友登山，但是由于还没有被开发，基础设施极为落后，社会救援也无法及时达到。蒲麦地村，是离牛背山顶最近的一个有人居住的小村落，村子基本呈现出中国西南地区传统的乡村面貌，坡屋顶、小青瓦，民风淳朴。正如中国大部分的偏远村庄一样，成年的劳动力大都在城市打工，村里更多的是留守儿童和空巢老人，很多村舍也是年久失修。资深义工大雁他们希望在这里建造一个给年轻人提供公益实践的基地，来帮助遇险的驴友的同时，也可以为村里的老人、儿童提供服务和帮助。为了保证公益实践的开支，他们也需要这个房子有一定的青年旅社的功能。

改造前的房子是一个传统的破旧民居，门前一个被当地叫作"坝子"的平台，木结构坡屋顶，但瓦已坏，平台上的首层空间被厚重的墙体分割成几个昏暗的小房间，屋顶阁楼已破旧不堪，没有厕所厨房，在坝子的南侧有一个后期农民自己加建的方正的砖房，与环境极不协调且不抗震。

我们的改造策略是在完善基本使用功能的前提下，让这个建筑更具有开放性与公共性，可为更多的人群服务，从建筑空间与结构上，在创新的同时，又具有中国传统建筑的记忆与灵魂，使其与村落、环境相协调，融为一体。

于是在一层我们保留并加固了内部的木结构，拆除了面向坝子的厚重墙体，以及内部隔墙，使一层完全开放，作为最重要的公共空间，可具有读书阅览、酒吧、会议等多种功能。重新设计的钢网架玻璃门，可以存储木柴，在完全打开的时候，将室内外融为一体。

坝子北侧的破旧猪圈被拆除，保留了木结构和坡屋顶，加建了围墙以及下水排污设施，改造为厨房和淋浴间以及卫生间，这也是整个村子唯一的一处有抽水马桶的卫生设施。

我们拆除了坝子南侧后建的砖房，还原了坝子原本的空间，并加建了一个木结构的构筑物，顶部覆瓦，可遮风避雨。在增大坝子使用率的同时，也形成了一个独特的观景平台。

在这个项目中，我们尽可能地使用当地村民作为主要的劳动力，用最常见、最基本的建筑材料和传统的搭建方式，比如当地石块的砌墙方式、坡屋顶与小青瓦的延续。在加建的构筑部分我们采用了数字化的设计方法与生成逻辑。

面对主屋你可以看到从左至右，逐渐是由传统转变到了现代，甚至是对未来的探索。一个和背后大山、云海相呼应的有机形态的屋顶呈现了出来。

内部看似是传统的木结构，但又是一种数字化的全新表现。这里所采用的材料是由四川本地盛产的慈竹所提炼而成的新型竹基纤维复合材料，具有高强度、耐潮湿、阻燃等特性、可循环再生、低碳环保。但对这种异形结构的加工，供应商也是首次尝试，结合建筑师的模型、图纸在施工现场放样加工，工厂预制与现场手工相结合。

主持建筑师李道德在解释这部分特殊的设计时，说道："这个起伏的屋顶与背后的大山以及远方的云海之间存在着形式上的某种关联，但我们更希望营造的是内心与情感上的联系，当'驴友'或者志愿者，甚至是村民们，徒步多时至此，远远可以看到村口有这么一个小小的独特而又熟悉的建筑泛着微微的暖光，就像是航船在大海航行中看到了灯塔，是给人们的一个召唤与鼓励，有着一种强烈的归属感。"

第六单元　建筑施工

·建筑施工简介·

施工就是将建筑设计转化为现实中的建筑物。施工必须严格遵照设计图纸，从而实现既定目标，并且在要求的时间内按预期的成本完成施工。

施工操作通常是根据专业领域来分类的。建筑施工过程可以分为以下几个阶段：

- 施工计划；
- 现场准备；
- 土方；
- 地基处理；
- 钢结构安装；
- 混凝土浇筑；
- 工地清理。

施工计划

施工计划开始于对建筑图纸和基本规范的详细研究。通过这一研究，准备好所有工作项目的清单。这些项目包括：材料、设备和劳动力的采购计划。这个阶段是最重要的，因为好的规划可以有效、经济地完成施工。每个阶段所需的建筑设备、劳动力和建筑材料都应适时地保证供应，因此，认真规划是必要的。

现场准备

"三通一平"确保施工工地通水、通电、通路，并在项目开始前平整土地。这包括除去和清理所有表面和场地上的生长物。

土方

土方包括开挖和填土。挖掘过程通常是以清除地表的有机土层开始的，这些土会被再利用来美化新建建筑的环境。挖掘可以由一些挖掘工具来完成，例如铲、拉铲挖掘机、抓斗、吊车和铲土机。填土以后，几乎都要压实以防止随后的沉降。

地基处理

当地下勘探显示结构要用到的基础范围内存在结构缺陷时,基础必须要加强。水的通路、洞穴、裂缝、断层和其他的缺陷都可以通过灌浆来填实和加强。灌浆是在压力下注入流体的混合物。流体随后在地层的空隙间凝固。大多数灌浆是用水泥和水的混合物做的,但也有其他的混合原料,如沥青、水泥、黏土和沉淀的化学物质。

钢结构安装

钢结构的施工包括卷钢厂钢材或工厂预制钢材的现场装配。钢材包括钢梁、钢柱或通过铆钉、螺钉、焊接连接而成的小型构件。

混凝土浇筑

混凝土施工包含以下几个操作:模板制作、混凝土生产、浇筑及养护。在可能的情况下,混凝土可以通过斜槽直接从搅拌车滑下进行浇筑,或者可以从起重机或架空索道控制的铲斗里倒下浇筑,还可以通过特殊混凝土泵浇筑于现场。

工地的清理

当施工完成时,工地必须彻底清理干净并美化,最后建筑物移交经理和维修人员。

·楼房设计中的安全因素·

在正常和紧急情况下,建筑物的设计应始终包括安全因素以防止故障。本节介绍了建筑物及其居住者如何抵御强风、地震、水、火和闪电的一般设计原则。

防风

在实际设计中,风和地震可以视为水平荷载或横向荷载。虽然风荷载和地震荷载可能具有垂直分量,但这些分量通常很小,很容易被柱子和承重墙抵抗。

在强烈地震或大风发生的可能性很小的地区,在建筑物内提供对两种荷载的相当大的抗力仍然是可取的。许多情况下,与忽略大风或抗震性的设计相比,这种阻力几乎或根本不增加成本。

防震保护

建筑物的设计应能承受几乎无处不在的轻微地震,而不造成破坏。这些原则要求避免倒塌,阻尼建筑物的振动,并尽量减少对结构和非结构部件的损坏。建筑物的抗震设计应当考虑到大的漂移,例如,通过在不需要刚性连接的相邻建筑构件之间提供间隙,并且允许这些构件滑动。因此,隔墙和窗户应该可以在框架中自由移动,这样当地震破坏框架时就不会发生损坏。

防水

无论是洪水冲撞或冲进建筑物,还是大雨灌入建筑物内部,从倾斜处泄漏,或是从外部围栏渗漏,水都会给建筑物造成巨大的损坏。

防护措施可分为两类:防洪和防水。防洪是指防止由河水溢出河岸而造成地表水泛滥;防水指防止地下水、雨水和融雪穿透建筑物的外围。

防火

消防工作有两个不同的方面：生命安全和财产保护。想要消除所有可燃材料或所有潜在的点火源是不可能的。因此，在大多数情况下尽管我们尽最大的努力防火，意外火灾仍会发生，我们必须制定适当的防火计划，采取措施来尽量减少火灾造成的损失。

设计者的首要义务是在提供客户需要的设施的同时满足法律要求，特别是适用的建筑规范要求。建筑规范包含消防安全要求，或者引用指定的一些公认的标准。如果没有其他原因，许多业主还要求咨询他们自己的保险承运人以获得最优惠的保险费率。

防雷

闪电是一种云和地面之间的高压、强烈放电现象，可能冲击和破坏任何雷暴发生区的生命和财产。然而，通过安装特殊的电气系统，建筑物及其居住者可以免受雷电危害。但由于不完备或不良安装可能造成比完全没有保护更严重的损害或伤害，因此，防雷系统应该由专家设计和安装。

作为建筑物所需的其他电气系统的补充，防雷系统增加了建筑物的建设成本。因此，房屋所有者必须决定潜在的损失是否值得增加支出。在这样做时，业主应比较保险成本与保护制度的成本，以弥补损失。

第七单元 房屋设备

·房屋设备简介·

在人们学会了建造有牢固的墙体、地面和屋顶的房屋之后，他们就开始发明更多的能使他们生活更方便、更舒适的方法。

其中的一个方法就是环境卫生工程，即提供清洁、安全的给水系统以及利用排水和下水道系统处置剩余水和废料。给水工程的主要设备就是两座水库：蓄水库和配水库，以及其间的管道。过滤、曝气、氯化、沉淀和活性污泥过程是最普通的使水净化的过程。污水处理厂就是将污水中的有害物除掉的工厂。污水工程主要指输送污水的管道或排水管，但它可以包括所有污水的收集和清除。

通风和空调被人们认为是一回事。空调的含义是通过净化、冷却、加温、干燥、湿润使提供给所用房屋的空气达到所要求的状态。我们说空气是通过这些过程来调解的。这样可以创造任何类型的空气环境，但要花费很多钱。通风是指供给新鲜空气，排除污浊气体和热气，以及为使空气冷却和清新而调节空气的流量。

照明（采光）在建筑物中对视觉和实用性都是必需的。住宅内通常有两种采光——总体采光和局部采光。总体采光应该是这样布置的，当我们分别打开各种不同的灯时，我们可以得到适应各种不同场合的预期效果：明亮的、中等亮度或较暗的灯光。局部照明是通过落地灯或管灯来实现的。

现代城市里的家庭很少看到用煤或柴火来烧饭、取暖，而是用天然气。地下和海底都有天然气，先把它引到地面上来，再输送到净化厂，然后有的直接用管道输送到住宅，有的制成液态石油气送往其他城市。

电梯是一种被安装在高层建筑物里方便省力的装置。借助于电梯，人和物可以高效率地、安全地在大楼内上下。如今电梯已经计算机化了。

更有甚者，就是住宅内的电器，诸如电视机、电话、计算机、录像机的发展。这些电器给人们在通信、工作和娱乐上带来了无法描述的便捷。目前，由建筑研究协会研制的多功能"精巧屋"也已在美国得到应用，其很多功能都是由计算机控制的。让我们来设想一下：拿起车上的手提电话，拨通家里，然后，只需按电话上的号码便可以吩咐屋子该做些什么……

这是家庭自动化的开端还是结局？

·铜陵山居·

传统性与当代性在现今中国的建筑实践中永远都处于一种在交织中的并行状态。就像铜陵山居这个项目，地处皖南一个僻静山村，项目起始于一幢徽州与沿江风格融合的普通民居。原有建筑处于全村最高的山顶，占地较小，而且房屋极为残破，已超过十年未有人居住，房屋四周皆为杂草和灌木覆盖。其东西向三跨，南北向一跨，原有屋面和墙体损毁严重。

故而从平面布局上我们考虑西向增加一跨，直抵一侧岩石山体，南北向部分增加一跨，以形成较宽阔的起居室空间。在平面上部分引入曲线，一个异化的虚体从原有投影轮廓中抽离出来。由于原有层高的限制和原有屋面完全破损不能再用，我们选择将原有建筑加高至两层，并结合平面上抽离出的双生虚形在空间上放样拉伸，并将前面的部分脊线压低，形成前后错落的空间连续曲面。传统的折面屋面和旁边抽离出来的流线型融合成一体，并暗合了中国文化"道生一，一生二"的宇宙观。整个屋面以青瓦覆盖，在鸟瞰角度形成独特的形态，与现有的古村落，既融合又出挑，并将室内空间的特征从外部进行首次表述。

如此形成了东西向的四跨空间。从东起，一跨形成地块南向的作为起居室的前厅空间；一跨与横向展开的部分残墙形成了庭院空间及其二层的玻璃观景平台；一跨与西侧的原建筑墙体形成了卧室空间；增加的一跨向西将山体和营造的部分景观空间纳入其半开放的檐下虚空间，由此在南立面上构成了一幅传统语境下的当代拼接画。

故其东侧立面在原有建筑的单脊面一侧又新生一脊，进一步暗示了这种双生关系的线索，并形成了一种宁静的平衡。新脊自成一跨，其中一半并入原建筑外轮廓，另一半形成檐下外廊，形成东立面上的虚实对比，并与南侧的下沉庭院景观形成对话。与南立面的横向展开的散点透视关系不同，东向悬崖方向的姿态更加倾向于一种单点透视的主体画面感，从中透射出的室内的铜饰构件及出挑的露台和挑檐，昭示着一种活力和期待感。

与外部廊道空间并置的，即为建筑的主要内部活动空间，由上述一系列新老墙体交错而成。故而原残损的墙体在建筑室内变成了数段内部墙体和室内隔断。从东向西，露台、起居室、餐厅、厨房、庭院、卧室由其私密性的逐级递减向西一字排开。居者通过外部廊道空间从室外台阶由南进入起居室，在此开展家庭聚会和活动。首层局部挑空，以获得更好的公共性和视野，从此向右到达半室外的露台部分远眺群山或俯瞰整个村落，从此向左到达更私密的餐饮空间和与之对应的厨房部分。建筑首层设有一间卧室及洗手间，从起居室的楼梯空间上去是阁楼上的两间卧室和洗手间。

中庭空间作为整个平面布局的核心部分，完成了由原建筑室内空间变成新建筑的室内空间的属性转换。整个中庭空间与西侧的檐下景观虚空间形成了东西一线的虚实关系的反复转换。配合悬浮的飘板楼梯，东侧的落地玻璃形成了传统民居空间不曾达到的居室空间的通透性。南北向则通过一系列线性高窗将双曲线屋面与墙体相分离，强化屋面的漂浮感和当代性，通过一系列小窗与楼梯主卧等重要建筑构件形成视线的呼应和室内外的对景关系。

为了快速实现曲线屋面形式，同时弱化结构构架尺寸对于需要保留的传统建筑元素和构件的影响，整个建筑采用了现场加工钢结构的建造方式。我们提取出平行于山墙面方向的若干"切片"式折线以及沿屋脊方向的几条结构曲线，形成整个屋面的钢结构空间网格体系。在这个结构体系下，屋顶的曲面形态可以通过其表面上的像素式的木挂瓦工艺在空间网格下的角度和位置进行更细微的调整和优化。基于钢结构的构架形式，原有墙体部分在建造过程中先由原墙体的老砖逐块标号保存，并待新基础及主体钢结构施工结束后原位复建。而复建砖墙的过程中，为了将钢结构掩藏于传统建造元素，许多老砖需要被切成薄片作为外贴饰面使用。新产生的墙体部分由当地其他建筑上的不同型号的老砖砌筑而成。新砌筑的部分刷上徽派建筑常见的白漆。并在两种砖体交接处进行了处理和做旧，使二者形成既和谐又统一的关系。整个建筑的立柱及屋面也均采用了其他老旧民居上回收的老料老瓦，并由当地工匠按本地传统工法砌筑而成。这样一方面在建构方式上回应了本土的文化性；另一方面也体现了可持续的生态理念。

第八单元　室内装饰

·绿色室内装饰·

众所周知，绿色室内装饰是一个新兴的、动态的和发展中的概念，它通过科技手段带来一种可持续发展的，实现人与自然和谐共存的理想模式。绿色室内装饰应尽可能地减少材料浪费引起的垃圾污染，充分利用自然光源和通风，同时通过绿色植物的合理布置，美化室内环境，提高空气质量。

在选择室内装饰材料时，需要根据建筑的类型、档次和使用部位的具体要求，来巧妙合理地运用材料的质感、线型和色彩，以便使建筑装饰满足一定的功能，适应一定的环境，发挥出最佳的装饰效果。

墙面材料

我们倾向于选用一些环保的墙纸，现在绿色设计与装饰追求返璞归真，选择运用天然植物装饰墙面并不是不无可能，例如现在棉、麻、藤制品等作为基材的天然墙饰。

地面材料

石材、木材、地毯、瓷砖、塑料地板是五种最常见的和不可或缺的地面材料。目前新出现的塑料地板正逐渐向环保靠近。塑料地板，表面光洁，耐老化性好，轻质高强，颜色丰富，施工方便，造价相对较低。微晶石作为一种天然无机材料为大众所接受。

自然光源

室内绿色装饰也可以通过改变窗口的位置来充分使用自然光源。如果需要，可使用透明的屋顶来提供更多良好的采光。充分利用自然光源可以避免不必要的资源浪费，节约能源，增加光度。

健康的色彩搭配

色彩运用得当不仅使视觉上感觉舒服，心理上也有一种潜移默化的影响。绿色空间就是需要这样人性化的色彩搭配，在居住空间融入正确的色彩搭配，不仅给入住业主的健康带来益处，也会使其身心得到放松。

空间的功能组合分布

合理的空间组织安排在绿色室内装饰中起到很重要的作用。所谓"合理的分布"，不仅要满足身心，美化外部空间，还要考虑到空间功能组织。室内设计的基本内容是如何利用空间。在建筑师和室内设计师创造和修改空间格局时，他们要与那些看到、使用和占有这些空间的人们交流理念和感觉。

·室内装饰设计的五个基本要素·

设计是以令人愉快的方式结合不同元素的艺术。装饰房屋的五个基本元素是颜色、纹理、线条、形状和空间。每种元素都有其独特的特点，一旦把它们综合运用，就会创造出别样的吸引力，从而美化房屋。了解这些元素，以及它们如何共同创造出合适的风格是很重要的。

颜色

当选择墙纸、油漆、窗帘、家具甚至地毯时，你首先要考虑的是颜色。颜色决定了房间的色调。它可以让房间看起来更大或更小，更温暖或更凉爽。主色调会给房间一个特定的情绪，而且来访者会立即注意到。室内设计中最重要的考虑因素之一是选择容易搭配的颜色。颜色可以用来吸引你的注意力，或者掩盖你最不喜欢的特征。在选择颜色时，尽量在同一家居中只选择几种颜色，或是互相衬托。当然，不要试图让它们完全搭配。这几乎是不可能的，但选择不同的色调要容易得多。

纹理

当为房间挑选物品时，接下来要注意的是材质。这在选择室内装潢和窗帘时尤其重要。粗犷、粗糙的纹理会创造出更休闲、温馨的感觉，而光滑、有光泽的质地，如丝绸，则让人感觉更正式。虽然你可以混合和搭配许多不同的面料，但是尽量坚持那些互补的面料。想要形成鲜明的对比，我们要避免使用不同质地，而应运用颜色等其他元素的优势。纹理也可以延伸到地板和墙壁。许多类型的墙纸都有非常鲜明的纹理，甚至油漆也会有不同程度的光泽。只有认真思量房间里的所有元素，才能创造出有凝聚力的外观和感觉。

线条和形状

接下来的两个室内设计元素经常被人们忽略，它们是形状和线条。在大多数房间里，根据墙面要求主要采用直线。然而，你也会发现其他家具的形状，以及墙壁或地板上的条纹和图案设计也都采用直线线条。垂直线可以创造高度，并给房间增添一点仪式感，而水平线比较安静，因此，垂直线和水平线相结合十分必要，可以创造出平衡感。斜线会吸引注意力，应该少用，因为它们会产生令人眼花缭乱的效果。曲线会使房间变得柔和，产生女性气质。这些线条创建出形状：室内装饰中最主要的形式是长方形，如沙发和咖啡桌；圆形会柔化房间刚性的感觉；三角形会感觉更稳定。

空间

只要空间使用得当，房间可以给人感觉更小或更大。明亮和较浅的颜色可能使房间显得宽敞，而暗和沉闷的颜色可能使房间显得局促。安置在墙上的家具使房间显得空间更大。而以不同的方式划分空间，一个房间便可能产生或大或小的错觉。

第九单元 园林设计

·景观设计风格·

人们在开始一项新景观项目之前要做的重要决定是选择一种最适合自己家的外观和感觉，同时还要体现自己个性的景观设计风格，这是实现房屋和花园之间统一的最佳方式。以下是各种景观设计风格的样本。

规整式

规整式的景观设计在很大程度上依赖直线和几何形状。植被被整齐有序地修剪以保持规整的效果。乔治王朝的花园所反映的风格显然属于此类。

非规整式

非规整式的景观设计与规整的风格正好相反。通过使用曲线和不规则形状来实现一种更为轻松的感觉。植被以随意的方式分布来创造出比较自然的景观。

托斯卡纳风格

托斯卡纳景观源自意大利南部的托斯卡纳地区。这种风格营造出一种古式的氛围，让人联想到意大利的乡村。而使用石头、旧砖、铁艺、厚重的木梁和正宗的托斯卡纳植物是该风格的典型特征。

地中海风格

地中海景观设计营造了流水潺潺和植被繁茂的景象，同时唤起人们与家人和朋友放松，并品尝由香草制成的美味佳肴的想法。地中海田园风格反映了欧洲南部悠闲的地中海文化并融合了优雅自然的细节之处。

英伦风格

传统的英式园林风格源于英国文化。其最引人注目的是芬芳的鲜花和郁郁葱葱的植物，以及在僻静的休息区和蜿蜒的走道上匍匐的藤蔓和雄伟的林荫树所体现的浪漫情调。

热带风情

一个精心设计的热带花园通常情况下种满了叶子巨大、花色鲜艳的植物，非常漂亮。花园后面布满高大葱郁的树木，形成繁茂的植物区。

亚洲风格

亚洲园林设计风格试图小规模地模仿大自然，而其中自然的随意性起着主导作用。东方园林往往把"风水"和"九宫"艺术融入其中，借此达到内心的平静和生命的平衡。

现代风格

现代景观设计已迅速普及流行。简洁的线条、大胆的图案和新材料的使用在这种新式风格里都起着重要作用，而大规模的植物和一些抽象的样式也较为常见。

沙漠风情

沙漠景观设计是在炎热干燥地区的富裕社区流行的花园样式，因此大力推荐种植一些需要很少甚至没有水都能蓬勃生长的本土植物。遮荫良好的户外起居室也是沙漠景观的一个组成部分，但要记住，重点是在午后骄阳下放松一定要看起来很自然。

上面的例子只是一些比较流行的园林设计风格。我们可以将两个或多个风格结合，创造出独特的新风格，达到融会贯通的效果。

·著名的住宅景观设计·

赫苏斯·加林德斯坡和博·卡萨尔斯广场

在改造之前，这里是一块位于市中心的布满岩石的斜坡，荒凉且一无是处。设计师将斜坡用不同材料的三角形截面重新包裹起来，向人们展示了这一奇特的地形。布满岩石的斜坡被改造成了一个连接元素：被三角形面板所包裹的岩石上形成了一个阶梯，将两片不同高度的社区连接起来，为斜坡周围的原有道路分散了人流。设计师利用项目的地势优势，将斜坡上方打造成了毕尔巴鄂市的观景平台。他们拆除了赫苏斯·加林德斯和博·卡萨尔斯大道的十字路口,并在那里移植了大片的菩提树。旧变电所被改造成了儿童游乐场,将平台整合成一个布局合理、绿树成荫的人造地貌。

景园辉庭

景园辉庭是一个高端住宅项目,以水景装饰为特点,具有独特的建筑装饰风格。项目的户外空间沿着两条轴线展开。其中一条主轴线从主楼一直延伸到景观平台,中间点缀着水景喷泉和精致的雕塑作品。这条轴线横跨整个园区的水平布局,甚至还一直延伸到建筑的表面,形成了一种统一的装饰风格。另一条轴线则划分出形式不一的水景设施,如戏水喷泉、儿童游泳池、带状水池和按摩浴缸。线性的游泳池是这条轴线的亮点,泳池的线条在按摩浴缸凉亭那里戛然而止。设计师巧妙地将具有现代异域风情的景观灯光、装饰艺术风格景观、喷泉和大理石雕塑结合在一起,营造了一个华丽的现代景观建筑。

北京时代尊府

北京时代尊府以现代中式皇家园林风情为特色,大气浩然而又精致华丽,成为皇城根下一道美丽的都市风景线。SED新西林景观国际以"现代中式皇家气派演绎都市生活"为设计理念,在延续城市历史文脉和整体城市风格的前提下,给景观设计赋予了新的生命,采撷中式皇家园林高超的布景、独特的施工工艺等优点,同时又不拘泥于传统,在造园布局上以错位的景观轴线来布置景观空间,注重理水、障景、借景、对景的运用,并有机地融入风水运势理论,凸显了现代中式皇家园林的大气浩然和现代都市生活的精致与活力,给居住者一个现代感强烈的中式皇家园林。前院、中庭、后院的整体布局,蜿蜒的园路,丰富的植物,便利合理的活动场地,高低起伏的微地形和富有变化的行走路线,都尽可能地营造一种自然、宜人的生活空间。

阿尔塔米拉庄园

阿尔塔米拉庄园的建筑设计和景观设计完美地结合在了一起,共同创造了一个规模庞大的项目。由于设计师采用了当地石材或是与当地石材类似的材料,并且栽种的植物100%都是加州本地的,项目和周围的环境异常和谐。项目的园艺设计不仅贴近自然,而且贴近海洋。随风飘动的灌木丛是蓝绿色的,仿佛庄园下方的海浪,灌木的飘动方向和海浪相平行,形成了同样的视觉韵律。正如海浪在靠近沙滩时渐渐变弱,还会撞到沙洲上,庄园周围的植物也渐渐变得低矮,中间还时不时地出现大片的沙地。在最靠近内陆的客房周围种植的多浆植物好像海胆、海星和珊瑚一样,让人想起了海潮池和海口。

第十单元 生态与建筑

·绿色建筑·

绿色建筑(也被称为绿色建筑物或可持续建筑物)是指在整个建筑的生命周期:即从

选址到设计、施工、运行维护、改造和拆除，既环保又节约资源的建筑结构和使用过程。这就要求设计团队、建筑师、工程师以及客户在所有项目阶段需要紧密合作。绿色建筑实践扩展并补充了古典建筑设计所关注的经济性、实用性、耐用性和舒适性。

高昂的能源成本、对环境问题的关注和与20世纪70年代盒状封闭式建筑相关的"病态建筑综合征"所引发的忧患共同推动了绿色建筑运动的兴起。绿色建筑汇集了繁多的实践、技术和技能，以减少并最终消除建筑物对环境和人类健康的影响。它往往强调利用可再生资源，例如，通过被动式太阳能、主动式太阳能和光伏发电设备，利用太阳光并通过利用植物和树木建造绿色屋顶、雨水花园以及减少雨水径流。许多其他技术也被应用到绿色建筑中，如使用低冲击力的建材或使用填充碎石或透水混凝土取代传统的混凝土或沥青，以提高地下水的补给。

尽管绿色建筑所采用的做法或技术都在不断演变，可能也有地区差异，但是各种做法采用的基本原则是一致的，包括：提高选址和结构设计效率、节能、节水、提高材料利用率、提升室内环境质量、运营和维护的优化，以及减少废物和有毒物质排放。绿色建筑的本质是优化其中的一个或多个原则。此外，在适当的协同设计下，单独的绿色建筑技术可以相互参照而产生更大的累积效应。

从美学角度上，绿色建筑或可持续建筑设计的理念是建筑物与周边的自然特征和资源和谐发展。在设计可持续建筑时有几个关键步骤：从当地资源中指定"绿色"建材，减少负载，优化系统，并生成即时可再生能源。

关于构建环境友好型建筑最为人诟病的问题是价格。光伏、新设备和现代技术往往花费更多的钱，而绿色建筑的倡导者认为这种建筑方式益处颇多。绿色建筑减少了长期能源消耗而节省了开支。以一间大办公室为例，绿色设计工艺与精妙技术两者的结合不仅能减少能源消耗和对环境的负面影响，而且能降低运营成本，创造一个更为愉快的工作环境，增进雇员的身体健康，提高生产率，减少法律责任以及提高地产价值和租赁收入。

·推板办公楼·

推板办公楼位于两片完全不同的城市肌理之间：北部是高密度的城市街区；南部的肌理较为松散，包含规划清晰直观的基础设施。基于办公和节能的要求，设计将成熟的节能技术应用到独立的办公楼层和室外空间——例如庭院、阳台和花园。建筑内部的三个核心筒和一座中央大堂为办公区域的灵活划分提供了最大的可能；可以在保持结构不变的情况下满足单个或多个租户的空间要求。

大楼建于一段旧的铁路路基之上，占地约4 000平方米。在地块限制范围内，建筑形态长约150米，宽约21米，从建筑中间的开口可以看到临街的一幢历史建筑。为了打造这扇城市之"窗"并增加这街区的特色，设计采用了将长板"推出"至断裂的造型，并将两处断口向南弯曲。"推"的动作也造成了楼板的变形，生成多层退台，可直达工作区和室外电梯。城市之"窗"为二层提供了一个巨大的平台露台。各层退台都布置着大型盆栽，为员工提供了怡人的休憩场所。

推板办公楼因而有了两幅面孔：以其较为冷静的面与巴黎北部的城市肌理进行对话，将更加活力的一面朝向南部的大街。建筑采用木质表皮。成排的开窗有如一条条起伏的缎带，为室内提供了最佳的自然光线和光照控制。遵循可持续发展的理念，建筑采用FSC认证木材，以减小木表皮对森林采伐的影响。

推板大楼正是一个高能效、经济现状和高品质建筑相结合的典范。通过制造一个特别的开口，我们保留了原有的视线，增加了建筑的价值，也表现出对周边住户的尊重，让"打开"这个概念在形态和情感上都具有多重的意义。

推板大楼是巴黎第一个"生态街区"（eco-quartier）计划中的首个落成项目。大楼屋顶的264块光伏板每年将产生90兆瓦的电力。建筑也将启用一个灰水循环系统。45%的加热水能源将由22块太阳热能集热器产生。建筑的南侧表皮以及切口面都安装了遮阳设施，与外部隔热以减少热桥效应。综合以上的成熟技术和设施，这座高能效建筑的年均耗能仅为46千瓦时/平方米，因而获得了BBC Effinengle 能源标签，并符合"巴黎市气候计划（Plan Climat de la ville de Paris）"制定的目标。

Appendix Ⅲ: 词汇表

New Words

A

aberrant [æˈberənt] adj. 离开正路的；与正确（或真实情况）相悖的 U9TB
absorb [əbˈzɔːb] v. 吸收 U2TA
absorption [əbˈzɔːpʃn] n. 吸收 U3TA
abstract [ˈæbstrækt] adj. 抽象的 U9TA
abundant [əˈbʌndənt] adj. 丰富的，充足的 U9TA
accommodate [əˈkɒmədeɪt] vt. 容纳；使适应；向……提供住处 U7TB
accumulation [əˌkjuːmjəˈleɪʃn] n. 积累；堆积物；累积量 U10TB
achieve [əˈtʃiːv] vt. 取得；实现 U5TA
acoustical [əˈkuːstɪk] adj. 听觉的 U3TA
activated-sludge [ˈæktɪveɪtslʌdʒ] n. 活性污泥 U7TA
adjoin [əˈdʒɔɪn] vt. & vi. 邻近；附加；接，贴连 U6TB
adopt [əˈdɒpt] vt. 采用，采取 U2TA
adornment [əˈdɔːnmənt] n. 装饰，装饰品 U8TA
aeration [eəˈreɪʃn] n. 曝气 U7TA
aesthetic [iːsˈθetɪk] adj. 美的，美学的 U10TA
aesthetically [iːsˈθetɪkli] adv. 审美地，美学观点上地 U5TB
affluent [ˈæfluənt] adj. 丰富的，丰裕的 U9TA
alloy [ˈælɔɪ] v. 合铸；铸成合金 U4TA
altar [ˈɔːltə(r)] n. 祭坛，圣坛 U2TA
aluminum [æljəˈmɪniəm；ˌæləˈmɪniəm] n. 铝 U4TA
amphitheater [ˈæmfɪθɪətə(r)] n. 竞技场；斗兽场 U1TB
analyse [ˈænəlaɪz] vt. 分析，研究 U3TA

· 165 ·

anchor [ˈæŋkə(r)] n. 锚；靠山 U9TB
anteroom [ˈæntiru:m] n. 接待室，前厅 U7TB
antisepsis [ˌæntɪˈsepsɪs] n. 防腐，消毒 U2TA
appliqué [əˈpli:kei] n. 贴花，嵌花，补花 U2TA
appose [əˈpoʊz] v. 并列，放……在对面 U7TB
aqueduct [ˈækwɪdʌkt] n. 渡槽，引水渠 U1TA
arch [ɑ:tʃ] n. 弓形（物） U2TB
architect [ˈɑ:kɪtekt] n. 建筑师，设计师 U3TB
architectural [ˌɑ:kɪˈtektʃərəl] adj. 建筑的；建筑学的 U2TA
architecturally [ˌɑ:kɪˈtektʃərəli] adv. 建筑上地 U5TA
architecture [ˈɑ:kɪtektʃə(r)] n. 建筑，建筑学 U2TA
arena [əˈri:nə] n. 舞台；竞技场 U1TB
array [əˈrei] n. 一串，一列 U9TA
asphalt [ˈæsfælt] n. 沥青，柏油 U6TA
assembly [əˈsembli] n. 装配；组装 U6TA
atop [əˈtɒp] adv.& prep. 在（……）顶上 U1TB
atrium [ˈeɪtrɪəm] n. (现代建筑物开阔的）中庭，天井；心房 U7TB
authentic [ɔ:ˈθentɪk] adj. 真正的 U9TA
axis [ˈæksɪs] n. 轴 U2TB

B

balcony [ˈbælkəni] n. 阳台；露台 U9TB
barren [ˈbærən] adj. 贫瘠的 U9TB
base [beɪs] n. 基础；基地；根据；vt. 基于；把……放在或设在（基地）上 U5TB
batten [ˈbætn] n. 板条；压条 vt. 用板条或压条固定 U7TB
beam [bi:m] n. 梁，栋梁；束；光线 U2TA
beget [bɪˈget] vt. 产生，引起 U7TB
binder [ˈbaɪndə(r)] n. 黏合剂 U4TA
biodegradable [ˌbaɪəʊdɪˈgreɪdəbl] adj. 能进行生物降解的 U4TA
blend [blend] vt. 混合；（使）调和；vi. 掺杂；结合；n. 混合；混合物 U5TB
bloom [blu:m] n. 最盛期，繁荣；vi. 大量出现 U1TA
bold [bəʊld] adj. 勇敢的，大胆的；U9TA
bolting [ˈbəʊltɪŋ] n. 螺栓连接 U6TA
boost [bu:st] vt. 促进，提高 U10TA
boulevard [ˈbu:ləvɑ:d] n. 大马路；林荫大道 U10TB
brick [brɪk] n. 砖 U2TA

brittle [ˈbrɪtl] adj.脆性的 U5TA
bucket [ˈbʌkɪt] n. 铲斗 U6TA
Buddhism [ˈbʊdɪzəm] n.佛教 U2TB

C

cableway [ˈkeɪbəlˌweɪ] n. 索道，缆道 U6TA
canal [kəˈnæl] n. 运河 U1TA
candidate [ˈkændɪdət] n.候选；候选人 U1TB
cany [ˈkeɪni] adj.藤的，藤制的 U8TA
cathedral [kəˈθiːdrəl] n. 大教堂 U1TA
cavity [ˈkævəti] n. 洞穴 U6TA
ceramic [səˈræmɪk] adj.陶瓷的 U8TA
certify [ˈsɜːtɪfaɪ] vt.（尤指书面）证明；发证书给…… U10TB
chlorination [ˌklɔːrɪˈneɪʃn] n. 氯化 U7TA
chute [ʃuːt] vt. 用斜槽或斜道运送 U6TA
circulation [ˌsɜːkjəˈleɪʃn] n. 活动空间，环流 U3TA
civilization [ˌsɪvəlaɪˈzeɪʃn] n.文明，文化 U2TA
civilize [ˈsɪvəlaɪz] v.使文明，教化 U2TA
clamshell [ˈklæmʃel] n. 抓岩机；抓斗 U6TA
classify [ˈklæsɪfaɪ] vt. 分类，归类 U6TA
clerestory [ˈklɪəstɔːri] n.天窗，通风窗 U7TB
coexistence [ˌkəʊɪɡˈzɪstəns] n.共存 U8TA
cohesive [kəʊˈhiːsɪv] a. 结成一个整体的 U8TB
cohesive [kəʊˈhiːsɪv] adj.有凝聚力的，紧密结合的 U1TA
collage [ˈkɒlɑːʒ] n.拼贴画；拼贴艺术；vt. 拼贴；vi. 制作拼贴 U7TB
collapse [kəˈlæps] vt.使倒塌，使坍塌 U2TA
collocation [ˌkɒləˈkeɪʃn] n.搭配，配置 U8TA
colossal [kəˈlɒsl] adj. 巨大的；庞大的 U1TA
colosseum [ˌkɒləˈsiːəm] n. 竞技场，斗兽场 U1TA
column [ˈkɒləm] n.柱，圆柱 U2TA
combination [ˌkɒmbɪˈneɪʃn] n.组合，结合 U8TA
combustible [kəmˈbʌstəbl] adj. 易燃的，可燃的 U6TB
commemorative [kəˈmemərətɪv] adj. 纪念的，纪念性的 U1TA
community [kəˈmjuːnəti] n. 社区；社会团体；共同体 U4TB
complement [ˈkɒmplɪment] vt.补足，补充 U10TA
compliment [ˈkɒmplɪmənt] vt. 向……道贺；称赞；向……致意 U8TB

composite [ˈkɒmpəzɪt] adj. 混合成的 U4TA
compression [kəmˈpreʃn] n. 压缩，紧缩 U4TA
compressive [kəmˈpresɪv] adj. 有压缩力的 U4TA
computerize [kəmˈpju:təraɪz] vt. （使）电子计算机化 U7TA
concept [ˈkɒnsept] n. 观念，概念 U8TA
concern [kənˈsɜ:n] vt. 对……关系，对……有重要性 U3TA
concrete [ˈkɒnkri:t] n. 混凝土 U4TA
condition [kənˈdɪʃn] vt. 支配，决定，限制 U3TA
configuration [kənˌfɪgəˈreɪʃn] n. 外形，形态 U5TA
constitute [ˈkɒnstɪtju:t] v. 构成，组成 U2TA
constrain [kənˈstreɪn] v. 约束，束缚 U5TA
contaminant [kənˈtæmɪnənt] n. 污染物，致污物 U7TA
contemporary [kənˈtemprəri] adj. 当代的，现代的 U3TB
continent [ˈkɒntɪnənt] n. 洲，大陆 U2TA
coral [ˈkɒrəl] n. 珊瑚 U9TB
cosmology [kɒzˈmɒlədʒi] n. 宇宙学；U7TB
coverage [ˈkʌvərɪdʒ] n. 覆盖 U3TA
craftsmanship [ˈkrɑ:ftsmənʃɪp] n. 手艺，精巧的技艺 U3TA
crane [kreɪn] n. 吊车，起重机 U6TA
critical [ˈkrɪtɪkl] adj. 决定性的 U4TA
cross-sectional [krɒsˈsekʃənl] n. 横截面 U5TA
crystal [ˈkrɪstl] adj. 透明的，清澈的 U9TA
cuisine [kwɪˈzi:n] n. 菜肴 U9TA
cumulative [ˈkju:mjələtɪv] adj. 累积的；渐增的 U10TA
curing [ˈkjʊə(r)ɪŋ] n. 固化，养护 U6TA
curve [kɜ:v] n. 弧线，曲线 U2TA

D

damp [dæmp] vi. [物]阻尼；减幅 U6TB
decline [dɪˈklaɪn] n. 下降；减少 U2TB
decorate [ˈdekəreɪt] vt. 装饰；点缀；粉刷；U8TB
decoration [ˌdekəˈreɪʃn] n. 装饰，装潢 U8TA
decorative [ˈdekərətɪv] adj. 装饰的 U1TA
dedicate [ˈdedɪkeɪt] vt. 奉献，献身 U5TB
deflect [dɪˈflekt] v. 倾斜，使弯曲 U5TA
deforestation [ˌdi:ˌfɒrɪˈsteɪʃn] n. 采伐森林，森林开伐 U10TB

deformability [dɪfɔ:məˈbɪlɪti] n.可变形性 U5TA
demolition [ˌdeməˈlɪʃn] n.毁坏，破坏，拆毁 U10TA
demonstrative [dɪˈmɒnstrətɪv] adj. 说明的，表明的 U1TB
dense [dens] adj. 密集的，稠密的；浓密的，浓厚的 U10TB
desperate [ˈdespərət] adj. 绝望的；铤而走险的；极度渴望的 U5TB
destructive [dɪˈstrʌktɪv] adj. 毁灭性的 U3TA
device [dɪˈvaɪs] n. 装置，装置物 U3TA
differentiate [ˌdɪfəˈrenʃɪeɪt] vt. 区别，区分，辨别 U3TA
differentiation [ˌdɪfəˌrenʃɪˈeɪʃn] n. 区分，分别，辨别 U3TA
digital [ˈdɪdʒɪtl] adj. 数字的；数据的 U5TB
dilapidated [dɪˈlæpɪdeɪtɪd] adj. 残破的，衰败的；破旧 U7TB
dimension [daɪˈmenʃn] n.维度，尺寸 U5TA
discharge [dɪsˈtʃɑ:dʒ] v. 放出；流出 U6TB
disguise [dɪsˈgaɪz] v. 隐藏，遮盖 U8TB
disposable [dɪˈspəʊzəbl] adj. 一次性的 U4TA
dissimilate [ˈdɪsɪmɪleɪt] v. (使) 变得不同 U7TB
distortion [dɪˈstɔ:ʃn] n.扭曲，变形；失真，畸变 U10TB
distribution [ˌdɪstrɪˈbju:ʃn] n.分布，分配 U8TA
dome [dəʊm] n. 圆屋顶 U1TA
dominant [ˈdɒmɪnənt] a. 高耸的；突出的 U8TB
dominate [ˈdɒmɪneɪt] adj. 支配的，占优势的 U9TA
dragline [ˈdræɡˌlaɪn] n. 牵引绳索；[机] 拉铲挖土机 U6TA
drainage [ˈdreɪnɪdʒ] n. 排水，放水 U7TA
drift [drɪft] v. 流动，漂流；浮现 n. 漂移 U5TB
ductility [dʌkˈtɪlɪti] n. 延性 U5TA
durability [ˌdjʊərəˈbɪlɪti] n. 耐久性；持久性 U1TA
dynamic [daɪˈnæmɪk] adj.动态的 U8TA
dynasty [ˈdɪnəsti] n.王朝，朝代 U2TA

E

earthmoving [ˈɜ:θˌmu:vɪŋ] n. 土方；土方工作 U6TA
eave [i:v] n. 屋檐 U7TB
eaves [i:vz] n. 屋檐 U3TA
ecological [ˌi:kəˈlɒdʒɪkl] adj. 生态（学）的 U4TA
edifices [ˈedɪfɪs] n. 大建筑物 U2TB
effective [ɪˈfektɪv] a. 有效的，奏效的 U3TA

efficiency [ɪˈfɪʃnsi] *n.* 效率，效能；实力，能力 U10TB
element [ˈelɪmənt] *n.* 元素，原件 U5TA
elevation [ˌelɪˈveɪʃn] *n.* [建]正视图，立视图；高地，高度 U7TB
eliminate [ɪˈlɪmɪneɪt] *vt.* 排除，消除 U10TA
embankment [ɪmˈbæŋkmənt] *n.* 路堤；筑堤 U10TB
embellishment [ɪmˈbelɪʃmənt] *n.* 装饰，修饰 U2TB
emergence [iˈmɜːdʒəns] *n.* 出现，发生 U8TA
emperor [ˈempərə(r)] *n.* 皇帝；君主 U2TB
emphasize [ˈemfəsaɪz] *v.* 强调 U2TA
empire [ˈempaɪə(r)] *n.* 帝国，帝国领土 U1TA
enclosure [ɪnˈkləʊʒə(r)] *n.* 圈占；围绕；U6TB
encourage [ɪnˈkʌrɪdʒ] *vt.* 激励，支持 U3TA
energizing [ˈenədʒaɪzɪŋ] *v.* 充满活力的 U3TB
enlarge [ɪnˈlɑːdʒ] *vt.& vi.* 扩大，放大；扩展，扩充 U4TB
enrich [ɪnˈrɪtʃ] *vt.* 使富裕，使富有 U2TB
equipment [ɪˈkwɪpmənt] *n.* 装备（品），设备 U3TA
erection [ɪˈrekʃn] *n.* 架设，安装 U6TA
essence [ˈesns] *n.* 本质，实质；精华，精髓 U7TB
estuary [ˈestʃuəri] *n.* （江河入海的）河口，河口湾；港湾 U9TB
evident [ˈevɪdənt] *adj.* 明显的，显然的 U3TA
evoke [ɪˈvəʊk] *v.* 唤起，引起 U9TA
evolve [iˈvɒlv] *vt.* 使发展；使进化 U10TA
excavation [ˌekskəˈveɪʃn] *n.* 挖掘；开挖 U6TB
excavator [ˈekskəveɪtə(r)] *n.* 挖掘机 U6TA
exemplary [ɪɡˈzempləri] *adj.* 典型的；示范的 U10TB
exotic [ɪɡˈzɒtɪk] *adj.* 异国的；外来的；吸引人的 U9TB
expenditure [ɪkˈspendɪtʃə(r)] *n.* 开销，经费 U3TA
expense [ɪkˈspens] *n.* 花费 U3TA
exploit [ɪkˈsplɔɪt] *vt.* 开采；开拓；利用（……为自己谋利）；剥削 U4TB
expressive [ɪkˈspresɪv] *adj.* 表现的 U3TA
exquisite [ɪkˈskwɪzɪt] *adj.* 精致的，优美的 U2TA

F

fabric [ˈfæbrɪk] *n.* 织物; 布; 构造; （建筑物的）结构 U8TB
fabricate [ˈfæbrɪkeɪt] *vt.* 制造；装配 U6TA
facade [fəˈsɑːd] *n.* 建筑物的正面；外表 U4TB

facilitate [fəˈsɪlɪteɪt] vt. 使便利 U3TA
facilitation [fəˌsɪlɪˈteɪʃn] n. 便利，便利设备 U3TA
facility [fəˈsɪləti] n. 设备 U7TA
fault [fɔːlt] n. 断层 U6TA
feature [ˈfiːtʃə(r)] n.特征，特点 U2TA
fenestella [ˌfenəˈstelə] n. 小窗，窗状壁龛 U7TB
feudalism [ˈfjuːdəlɪzəm] n. 封建制度 U2TB
fiber [ˈfaɪbə(r)] n.光纤；（织物的）质地；纤维，纤维物质 U5TB
filtration [fɪlˈtreɪʃn] n. 过滤 U7TA
fissure [ˈfɪʃə(r)] n. 裂缝，裂隙 U6TA
flair [fleə(r)] n. 天资；天分 U1TA
flexibility [ˌfleksəˈbɪləti] n. 机动性，灵活性 U1TA
flood [flʌd] n. 水灾 U3TA
floodproofing [flʌdˈpruːfɪŋ] 防洪 U6TB
flush [flʌʃ] vi.奔流；冲刷 vt.（以水）冲刷，冲洗；n. 奔流，涌出 U5TB
foliage [ˈfəʊliɪdʒ] n. 树叶，植物 U9TA
foreshadow [fɔːˈʃædəʊ] vt. 预示，预兆 U1TA
formality [fɔːˈmæləti] n. 礼节;拘谨; 正式手续 U8TB
fortification [ˌfɔːtɪfɪˈkeɪʃn] n. 筑垒；防御工事 U1TB
foundation [faʊnˈdeɪʃn] n. 地基；基础 U6TA
fragment [ˈfræɡmənt] n. 碎片；片段；v.（使）碎裂，破裂 U5TB
fragrance [ˈfreɪɡrəns] n. 芳香；香气；香水(常用于广告语) U4TB
fragrant [ˈfreɪɡrənt] adj. 芬芳的，香的 U9TA
frame [freɪm] n. 框架 U5TA
freshen [ˈfreʃn] vt. 使新鲜；使清爽 U7TA
furnishing [ˈfɜːnɪʃɪŋ] n. 家具与陈设品（常用复数） U3TA

G

gable [ˈɡeɪbl] wall 山墙，山墙墙身，前脸墙 U7TB
generate [ˈdʒenəreɪt] vt. 形成，造成；产生（后代）；引起 U10TB
geometric [ˌdʒiːəˈmetrɪk] adj.几何图形的 U3TB
giant [ˈdʒaɪənt] adj.巨大的 U5TA
girder [ˈɡɜːdə(r)] n. 主梁，纵梁 U5TA
glaze [ɡleɪz] v.给……上釉，使光滑，使光亮 U2TA
glittering [ˈɡlɪtərɪŋ] adj. 辉煌的；光辉灿烂的 U1TB
gravel [ˈɡrævl] n. 沙砾，碎石； U4TA

grid [grɪd] *n.* 格子；（输电线路、天然气管道等的）系统网络 U10TB
groin [grɔɪn] *n.* 交叉拱 U1TA
grout [graʊt] *vt.* 灌浆 U6TA

H

habitable [ˈhæbɪtəbl] *adj.* 适于居住的 U3TA
harmonious [hɑːˈməʊnɪəs] *adj.* 和谐的，融洽的；悦耳的 U5TB/U1TA
harmonize [ˈhɑːmənaɪz] *vt.* 调和，和谐 U3TA
Hellenistic [ˌheliˈnɪstɪk] *adj.* 希腊风格的，希腊文化的 U1TB
hindrance [ˈhɪndrəns] *n.* 妨碍的人或物 U3TA
hollow [ˈhɒləʊ] *adj.* 中空的 U5TA
horizontal [ˌhɒrɪˈzɒntl] *adj.* 水平的 U6TB
horizontal [ˌhɒrɪˈzɒntl] *adj.* 水平的，横向的 U5TA
humidify [hjuːˈmɪdəfaɪ] *vt.* 使湿润；使潮湿 U7TA
hyperbola [haɪˈpɜːbələ] *n.* 双曲线 U7TB

I

illusion [ɪˈluːʒn] *n.* 错觉；幻想；错误观念 U8TB
illustrate [ˈɪləstreɪt] *v.i.*（用图，实例等）说明，阐明(+with)；U2TB
imaginative [ɪˈmædʒɪnətɪv] *adj.* 富于想象力的 U3TB
imitate [ˈɪmɪteɪt] *vt.* 模仿，效仿 U1TA
imperial [ɪmˈpɪərɪəl] *adj.* 帝国的，皇帝的 U2TA
imperialism [ɪmˈpɪərɪəlɪzəm] *n.* 帝国主义 U2TB
imperishable [ɪmˈperɪʃəbl] *adj.* 不朽的，不会腐烂的 U4TA
implement [ˈɪmplɪment] *vt.* 实施，执行；使生效，实现 U5TB
inaugurate [ɪˈnɔːgjəreɪt] *vt.* 开创；举行典礼 U1TB
incorporate [ɪnˈkɔːpəreɪt] *v.* 使混合或合并 U9TA
indigenous [ɪnˈdɪdʒənəs] *adj.* 土生土长的；生来的，固有的 U4TA
indispensable [ˌɪndɪˈspensəbl] *adj.* 不可缺少的 U8TA
inevitably [ɪnˈevɪtəbli] *adv.* 不可避免地，必然发生地 U3TA
infrastructure [ˈɪnfrəstrʌktʃə(r)] *n.* 基础设施；基础建设 U5TB/U10TB
inhabit [ɪnˈhæbɪt] *vt.* 居住；在……出现；填满；*vi.* 居住 U5TB/U7TB
injection [ɪnˈdʒekʃn] *n.* 注入 U6TA
innovation [ˌɪnəˈveɪʃn] *n.* 改革，创新 U1TA
innumerable [ɪˈnjuːmərəbl] *adj.* 无数的 U2TB
inspire [ɪnˈspaɪə(r)] *vt.* 鼓舞；激励 U1TB

install [ɪnˈstɔːl] vt. 安装 U7TA
instill [ɪnˈstɪl] vt. 逐渐持续地引入，灌输 U9TA
insulate [ˈɪnsjuleɪt] vt. 隔热，绝缘 U3TA
integral [ˈɪntɪgrəl] adj. 完整的 U2TB
integrate [ˈɪntɪgreɪt] v. 合并；成为一体；加入；融入群体 U7TB
intensify [ɪnˈtensɪfaɪ] vt.& vi. （使）增强，（使）加剧 U7TB
intervention [ˌɪntəˈvenʃn] n. 介入；调停；妨碍 U9TB
intimacy [ˈɪntɪməsi] n. 亲密；亲近 U1TA
intricate [ˈɪntrɪkət] adj. 错综复杂的；难理解的；曲折 U4TB
involve [ɪnˈvɒlv] vt. 包含，涉及 U3TA
irresistible [ˌɪrɪˈzɪstəbl] adj. 无法抗拒的；不可阻挡的 U1TB

J

joist [dʒɔɪst] n. 托梁，搁栅 U2TA

L

landscape [ˈlændskeɪp] vt. 对……做景观美化，给……做园林美化 U6TA
legionnaire [ˌliːdʒəˈneə(r)] n. 军团的士兵 U1TB
level [ˈlevl] vt. 平整，弄平 U6TA
liability [ˌlaɪəˈbɪləti] n. 责任；倾向 U10TA
lightweight [ˈlaɪtweɪt] n. 轻质；轻量级 U4TA
linearity [ˌlɪniˈærəti] n. 线型 U8TA
linen [ˈlɪnɪn] adj. 亚麻的，亚麻制品的 U8TA
lintel [ˈlɪntl] n. 过梁，（门或窗的）楣 U2TA
load [ləʊd] n. 负荷；负担；装载 U6TB
lush [lʌʃ] adj. 青葱的，繁茂的 U9TA

M

magnesium [mæɡˈniːziəm] n. [化]镁（金属元素） U4TA
magnificent [mæɡˈnɪfɪsnt] adj. 宏伟的 U2TB
maintenance [ˈmeɪntənəns] n. 维持，保持；保养，保管；维护；维修 U5TB
majority [məˈdʒɒrəti] n. 多数；（获胜的）票数；成年；法定年龄 U5TB
masonry [ˈmeɪsənri] n. 砌石；砌体 U4TA
mausoleum [ˌmɔːsəˈliːəm] n. 陵墓 U1TB
maximize [ˈmæksɪmaɪz] vt. 最大化，使（某事物）增至最大限度 U5TB
Mayan [ˈmɑːjən] adj. 玛雅人的，玛雅语的 n. 玛雅人，玛雅语 U1TB

meandering [miˈændə(r)ɪŋ] adj. 蜿蜒的，曲折的 U9TA
measurements [ˈmeʒəmənt] n. 尺度，尺寸 U3TA
metope [ˈmetoʊp] n. 墙面 U8TA
mill [mɪl] n. 工厂 U6TA
mimic [ˈmɪmɪk] vt. 模仿，模拟 U9TA
miracle [ˈmɪrəkl] n. 奇迹，圣迹 U2TB
modify [ˈmɒdɪfaɪ] vt. 减轻，缓和 U3TA
moisture [ˈmɔɪstʃə(r)] n. 水分；湿气；潮湿；降雨量 U5TB/U3TA
mold [məʊld] vt. 浇铸；塑造 U4TA
monastery [ˈmɒnəstri] n. 修道院，寺院 U1TB
monolithic [ˌmɒnəˈlɪθɪk] adj. 整体的 U5TA
monument [ˈmɒnjumənt] n. 纪念碑；遗迹；遗址 U1TB
motifs [məʊˈtiːf] n. 装饰图案 U2TB
muscular [ˈmʌskjələ(r)] adj. 肌肉的 U3TA

N

naturalistic [ˌnætʃrəˈlɪstɪk] adj. 自然的 U9TA
negligible [ˈneɡlɪdʒəbl] adj. 可以忽略的；微不足道的 U4TA
nondescript [ˈnɒndɪskrɪpt] adj. 难以描述的 U7TA
nubby [ˈnʌbɪ] adj. 有节的，块状的 U8TB

O

obligation [ˌɒblɪˈɡeɪʃn] n. 义务，责任 U6TB
occupant [ˈɒkjəpənt] n. 占有人；居住者 U6TB
octagon [ˈɒktəɡən] n. 八边形 U2TB
octagonal [ɒkˈtæɡənl] adj. 八边形的 U1TA
optimize [ˈɒptɪmaɪz] vt. 使最优化，使尽可能有效 U10TA
orderly [ˈɔːdəli] adj. 有秩序的，整齐的 U9TA
organic [ɔːˈɡænɪk] adj. 有机（体）的；有组织的，系统的 U5TB
orientation [ˌɔːriənˈteɪʃn] n. 朝向，认识环境 U3TA
oscillation [ˌɒsɪˈleɪʃn] n. 振动；波动；<物>振荡 U6TB
oval [ˈəʊvl] adj. 椭圆形的 U1TA
overhang [ˌəʊvəˈhæŋ] vt&vi (hung)悬于……之上；悬垂 U3TA

P

pagodas [pəˈɡəʊdə] n. 宝塔 U2TB

palette [ˈpælət] n. 调色板，颜料 U9TB
palette [ˈpælət] n. 主要色彩，主色调 U8TB
panel [ˈpænl] n. 镶板；面；（门、墙等上面的）嵌板 vt. 把……镶入框架内 U10TB
park [pɑːk] n. 停车场 U3TB
parlor [ˈpɑːlə(r)] n. 客厅；起居室；（旅馆中的）休息室 U7TB
particularize [pəˈtɪkjələraɪz] vt. 特别指出，具体化 U3TA
partition [pɑːˈtɪʃn] n. 隔离物；隔墙 U6TB
patio [ˈpætiəʊ] n. 露台，平台 U10TB
patron [ˈpeɪtrən] n. 赞助人，资助人 U1TA
pavilion [pəˈvɪliən] n. 亭，阁楼；展区，展馆 U2TA
pedestrian [pəˈdestriən] n. 步行者；行人 U9TB
penetrate [ˈpenətreɪt] vt. 穿透，刺入；渗入 U6TB
perform [pəˈfɔːm] vt. 执行，履行，做 U3TA
permeable [ˈpɜːmiəbl] adj. 可渗透的，具渗透性的 U10TA
perspective [pəˈspektɪv] n. 远景，景色；观点，看法；洞察力 U7TB
phase [feɪz] n. 方面，片段，阶段 U3TA
philosophy [fəˈlɒsəfi] n. 哲学 U2TA
photo-voltaic 聚光太阳电池方阵 U10TB
pillar [ˈpɪlə(r)] n. 柱子，支柱 U2TA
pitched [pɪtʃt] adj. （屋顶）有坡度的；v. 投（pitch的过去式和过去分词）；用沥青涂；排列 U5TB
pixel [ˈpɪksl] n. （显示器或电视机图像的）像素 U7TB
plastic [ˈplæstɪk] adj. 塑料的 U8TA
plate [pleɪt] n. 金属板 U4TA
playful [ˈpleɪfl] adj. 有趣的 U3TB
positivity [ˌpɒzəˈtɪvəti] n. 积极性 U4TB
postmodernism [ˌpəʊstˈmɒdənɪzəm] n. 后现代主义 U1TA
potentiality [pəˌtenʃiˈæləti] n. 可能性，潜力 U3TA
practical [ˈpræktɪkl] adj. 实际的 U3TA
precipitate [prɪˈsɪpɪteɪt] vt.& vi. [化]（使）沉淀 U6TA
predicted [prɪˈdɪktɪd] adj. 预测的；预期的 U6TA
predominate [prɪˈdɒmɪneɪt] vi. 占支配地位 U1TA
prefabricated [ˌpriːˈfæbrɪkeɪtɪd] adj. （建筑物）预制构件的 U4TA
preserve [prɪˈzɜːv] vt. 保护 U3TA
prevail [prɪˈveɪl] vi. 流行，盛行；获胜，占优势；说服，劝说 U7TB/U2TB
principle [ˈprɪnsəpl] n. 原则，原理；准则 U1TA

procurement [prəˈkjʊəmənt] n. 采购；采购信息 U6TA
profound [prəˈfaʊnd] adj. 深厚的，意义深远的 U2TA
proponent [prəˈpəʊnənt] n. 支持者，拥护者 U10TA
proportion [prəˈpɔːʃn] n. 比，比率 U1TA
prototype [ˈprəʊtətaɪp] n. 原型，雏形 U1TA
prune [pruːn] v. 修剪 U9TA
psychological [ˌsaɪkəˈlɒdʒɪkl] adj. 心理的，心理学的 U8TA
pump [pʌmp] n. 泵 vt. 用泵抽 U6TA
punctuate [ˈpʌŋktʃueɪt] vt. 加强，强调 U9TB
puppet [ˈpʌpɪt] n. 木偶；傀儡；受他人操纵的人 U4TB
purify [ˈpjʊərɪfaɪ] vt. 净化 U7TA

Q

quality [ˈkwɒləti] n. 品质 U3TA

R

recluse [rɪˈkluːs] n. 隐居者，遁世者，隐士 U7TB
rectangular [rekˈtæŋɡjələ(r)] adj. 长方，矩形的；成直角的 U10TB
recyclable [ˌriːˈsaɪkləbl] adj. 可循环再用的 U5TB
redefine [ˌriːdɪˈfaɪn] vt. 重新定义 U3TB
reflect [rɪˈflekt] vt. 反映 U2TA
reflex [ˈriːfleks] n. 反射，反射动作 U3TA
reinforce [ˌriːɪnˈfɔːs] vt. 加固；使更结实；加强 U4TA
relic [ˈrelɪk] n. 遗物，遗迹；废墟；纪念物 U7TB
reminiscent [ˌremɪˈnɪsnt] adj. 怀旧的，引起回忆往事的 U9TA
removal [rɪˈmuːvl] n. 清除 U7TA
renovation [ˌrenəˈveɪʃn] n. 翻新；修复 U10TA
repel [rɪˈpel] vt. 逐退，驱开 U3TA
replenishment [rɪˈplenɪʃmənt] n. 补给，补充 U10TA
reservoir [ˈrezəvwɑː(r)] n. 蓄水池；水库 U7TA
residential [ˌrezɪˈdenʃl] adj. 住宅的 U9TB
resin [ˈrezɪn] n. 树脂；合成树脂 U4TA
resistance [rɪˈzɪstəns] n. 阻力，抵抗 U5TA
rhythmic [ˈrɪðmɪk] adj. 有韵律的，有节奏的；格调优美的 U10TB
ridge [rɪdʒ] n. 背脊，峰；山脊，vt. 使成脊状，使隆起；vi. 使成脊状 U7TB
riveting [ˈrɪvɪtɪŋ] n. 铆接（法） U6TA

rotate [rəʊˈteɪt] vt.& vi. （使某物）旋转；使转动；使轮流，轮换 U4TB

S

sanitary [ˈsænətri] adj. 卫生的；清洁的 U7TA
savory [ˈseɪvəri] adj. 可口的，味美的 U9TA
scraper [ˈskreɪpə(r)] n. 铲土机 U6TA
sculptor [ˈskʌlptə(r)] n. 雕刻家 U3TB
sculpture [ˈskʌlptʃə(r)] n. 雕刻 U2TB
secluded [sɪˈkluːdɪd] adj. 僻静的，隐蔽的 U9TA
secular [ˈsekjələ(r)] adj. 非宗教的；俗界的 U4TA
sedimentation [ˌsedɪmenˈteɪʃn] n. 沉淀，沉降 U7TA
seek [siːk] vt. 试图，努力 U3TA
seismic [ˈsaɪzmɪk] adj. 地震的；由地震引起的 U6TB
settlement [ˈsetlmənt] n. 沉降 U6TA
sewage [ˈsuːɪdʒ] n. 污水；废水 U7TA
sewer [ˈsuːə(r)] n. 污水管，下水道 U7TA
sewerage [ˈsuːərɪdʒ] n. 污水工程，污物处理（系统）U7TA
shabby [ˈʃæbi] adj. 破旧的；卑鄙的；悭吝的 U5TB
sheet [ʃiːt] n. （金属材料）制成的薄板 U4TA
shelter [ˈʃeltə(r)] n. 居所；避难所 U3TB
shimmer [ˈʃɪmə(r)] vi. 闪烁发微光 n. 微光；闪光 U5TB
shovel [ˈʃʌvl] n. 铲子，铁锹 U6TA
shrub [ʃrʌb] n. 灌木；灌木丛 U9TB
silicon [ˈsɪlɪkən] n. <化>硅；硅元素 U4TA
site [saɪt] n. 现场 U6TA
skyscraper [ˈskaɪskreɪpə(r)] n. 摩天大楼 U5TA
slab [slæb] n. 厚板，平板 U4TA
slope [sləʊp] n. 斜坡 U9TB
smallpox [ˈsmɔːlpɒks] n. 天花 U1TB
solid [ˈsɒlɪd] adj. 结实的，坚固的 U5TA
solidify [səˈlɪdɪfaɪ] vt.& vi. 使凝固，固化 U6TA
spacious [ˈspeɪʃəs] adj. 宽敞的；广阔的 U7TB
span [spæn] n. 跨度，墩距；一段时间 U7TB
spatial [ˈspeɪʃl] adj. 空间的；存在于空间的；受空间条件限制的 U7TB
species [ˈspiːʃiːz] n. 物种；种类 U4TA
specification [ˌspesɪfɪˈkeɪʃn] n. 规格；规范 U6TA

spectacular [spek'tækjələ(r)] adj. 壮观的 U4TA

Sphinx [sfɪŋks] n. [埃]狮身人面 U1TA

spire ['spaɪə(r)] n. 塔尖；尖顶 U1TA

spiritual ['spɪrɪtʃuəl] adj. 精神的；心灵的；高尚的 U5TB

stagger ['stægə(r)] vi. 蹒跚；犹豫；动摇；vt. 使蹒跚，使摇摆 U7TB

standardization [ˌstændədaɪ'zeɪʃn] n. 标准化 U3TA

stiffness ['stɪfnəs] n. 硬度 U5TA

strata ['strɑːtə] n. 地层；岩层(stratum的名词复数) U6TA

strategy ['strætədʒi] n. 策略，战略；战略学 U5TB

stretch [stretʃ] v. 伸展；延伸；持续n. 伸展；弹性 adj. 可伸缩的；弹性的 U7TB

striking ['straɪkɪŋ] adj. 显著的，突出的 U3TB

stripping ['strɪpɪŋ] n. 清除，脱模 U6TA

structure ['strʌktʃə(r)] n. 结构 U5TA

subject ['sʌbdʒɪkt] adj. 受制于的（与to连用） U3TA

substance ['sʌbstəns] n. 物质，材料 U4TA

subsurface ['sʌbˌsɜːfəs] adj. 表面下的，地下的 U6TA

subtle ['sʌtl] adj. 微妙的，精细的 U8TA

succulent ['sʌkjələnt] n. 肉质植物；多汁植物 U9TB

sunken ['sʌŋkən] adj. 沉没的；凹陷的，下陷的 U7TB

suspend [sə'spend] v. 暂停；延缓；悬挂 U7TB

sustainable [sə'steɪnəbl] adj. 可持续的；可支撑的 U7TB

symmetry ['sɪmətri] n. 对称；对称美 U1TA

synergistic [ˌsɪnə'dʒɪstɪk] adj. 增效的，协作的，互相作用[促进]的 U10TA

synthetic [sɪn'θetɪk] adj. 合成的；人造的 U4TA

T

tectonic [tek'tɒnɪk] adj. 构造的，建筑的 U7TB

tensile ['tensaɪl] adj. 拉力的；张力的 U4TA

tension ['tenʃn] n. [物]张力，拉力 U4TA

terminus ['tɜːmɪnəs] n. 终点；终点站 U3TB

terrace ['terəs] n. 台阶，阶地；阳台；柱廊，门廊；斜坡上房屋间的街巷 U10TB

territory ['terətri] n. 领土，范围 U2TA

testify ['testɪfaɪ] vi. 证明；证实；作证 vt. 作证；证明，为……提供证明 U4TB

texture ['tekstʃə(r)] n. 质地，纹理，组织 U3TA

tile [taɪl] n. 瓦，瓷砖 U2TA

timber ['tɪmbə(r)] n. 木材，木料 U4TA

titanium [tɪˈteɪnɪəm] n. 钛 U3TB

topography [təˈpɒɡrəfi] n. 地形学；地形测量学 地貌 U9TB

transition [trænˈzɪʃn] n. 过渡，转变，变迁 U1TA

transitional [trænˈzɪʃənl；trænˈsɪʃnəl] adj. 变迁的，过渡期的；转变的；转移的 U7TB

transparency [trænsˈpærənsi] n. 透明；透明度；透明的东西 U4TB

transparent [trænsˈpærənt] adj. 透明的 U8TA

treatment [ˈtriːtmənt] n. 处理 U6TA

tropical [ˈtrɒpɪkl] adj. 热带的 U9TA

truss [trʌs] n. 构件，构架 U6TA

tube [tjuːb] n. 筒体 U5TA

tunnel [ˈtʌnl] n. 隧道；地道 U1TB

twig [twɪɡ] n. 细枝，嫩枝 U4TA

U

ultimately [ˈʌltɪmətli] adv. 最后，终极的 U3TA

unencumbered [ˌʌnɪnˈkʌmbəd] adj. 没有阻碍的，不受妨碍的；无负担的 U7TB

upholstery [ʌpˈhəʊlstəri] n. 家具装饰业；室内装饰品 U8TB

urchin [ˈɜːtʃɪn] n. 海胆 U9TB

utilitarian [ˌjuːtɪlɪˈteərɪən] adj. 实用的，以实用为主的，功利的 U3TA

utility [juːˈtɪləti] n. 功用，效用 U10TA

V

vegetation [ˌvedʒəˈteɪʃn] n. 植被，植物 U9TA

ventilation [ˌventɪˈleɪʃn] n. 通风 U7TA

veranda [vəˈrændə] n. 阳台，走廊 U7TB

version [ˈvɜːʃn] n. 版本；译文 U3TB

vertical [ˈvɜːtɪkl] adj. 垂直的，竖立的；n. [建]竖杆；垂直线，垂直面 U4TB

vertically [ˈvɜːtɪkli] adv. 垂直地；直立地 U9TB

virtually [ˈvɜːtʃuəli] adv. 实际上，事实上 U1TB

W

waterproofing [ˈwɔːtəpruːf] 防水 U6TB

welding [weldɪŋ] n. 焊接法，定位焊接 U6TA

welfare [ˈwelfeə(r)] n. 福利；幸福；繁荣；安宁 U5TB

· 179 ·

Y

youth hostel ['juːθ hɒstl] 青年招待所青年（学生）宿舍 U5TB

Phrases & Expressions

A

a thousand doors and ten thousand windows 千门之窗 U2TB
a series of 一系列；一连串 U7TB
a string of 一系列；一批；一连串 U7TB
according to 依照 U2TA
account for 说明，占，解决，得分 U2TA
adapt to 适用于 U4TA
air motion 空气流动 U7TA
air quality 空气质量 U8TA
apply in/to 运用于，应用于 U8TA
aromatic herb 草本香料植物 U9TA
art-deco landscape style 装饰艺术风格的景观 U9TB
as is known to all 众所周知 U8TA
at once 同时 U3TA

B

be based on... 建立在……基础上 U2TA
be combined with 与……联合 U5TA
be composed of 由……组成 U5TB
be concerned with 对……有关 U3TA
be conditioned by 受……限制 U3TA
be connected to 与……有联系，与……有关联 U3TB
be covered with 覆盖着 U3TB
be embedded in 嵌入 U5TA
be made from 由……所做成 U3TB
be regarded as 视为，被认为是 U1TB
be responsible for 对……负责 U5TA
be subject to 受控制于…… U3TA

· 180 ·

be traced back 追溯到…… U5TA
bearing wall 承重墙 U6TB
boundary beam 边梁 U5TA
building code 建筑规范 U6TB
by means of 依靠，凭借 U7TA

C

call for 呼吁 U3TB
call on 号召；要求；拜访（某人） U4TB
cast iron 铸铁 U1TA
Central Asian culture 中亚文化 U2TB
close up 维护 U5TA
coincide with 与……一致 U7TB
comply with 遵从，服从 U2TA
compressive capacity 抗压能力 U5TA
concrete structure 混凝土结构 U5TA
concrete placement 混凝土浇筑 U6TA
construction plans 建筑图纸 U6TA
constructional column 构造柱 U5TA
contribute to 捐献；促成；有助于 U10TB
coordinate to 与……协调 U5TB
corrosion resistance 耐腐蚀性 U4TA
cross T beam T形横截面 U5TA
curtain-wall construction 幕墙结构 U4TA
curved line 曲线 U8TB

D

dated back to 追溯至 U2TB
dead mass 自重 U5TA
derive from 由……起源；取自 U10TA
diagonal line 斜线，对角线 U8TB
distribution reservoir 配水库 U7TA
divided into 被合成 U2TB
drip down 滴下来 U7TB

E

earth fill 填方，填土 U6TA
economic accounting 经济核算 U3TA
electric power supplies 电力供应 U6TA
environmental design 环境设计 U3TA
environmental protection 环境保护 U8TA
environmental sanitary engineering 环境卫生工程 U7TA

F

floor plan 地面路线图；楼面布置图；楼面图 U9TB
foreign-style house 洋房 U2TB
forge ahead 稳步前进，开拓进取 U2TA
formal Style 规整式 U9TA
framed structure 框架结构 U5TA

G

gain ground 普及；发展；U9TA
general lighting 总体采光 U7TA
growth hormone secretion deficiency 生长激素分泌不足 U4TB

H

hand over 移交，交出 U6TA
horizontal line 水平线 U8TB
horizontal masonry wall 横墙 U5TA
horizontal plane 水平面 U9TB

I

ignition sources 火源 U6TB
in a sense 在某种意义上 U2TA
in addition to 除了 U8TA
in cooperation with 与……合作 U10TB
in harmony with 与……协调，与……一致 U10TA
in need of 需要…… U8TA
in order to 为了 U2TB
in popularity 流行 U9TA

· 182 ·

informal Style 非规整式 U9TA
insurance rate 保险费率 U6TB
interior decoration 室内装饰 U8TA
interior design 室内设计 U8TB

J

joint motor 马达 U5TA
jump-start 起动，发动 U10TA

L

lag behind 落后，退步 U5TB
landscape design 园林设计 U9TA
lightning protection system 防雷系统 U6TB
lime tree 椴树/菩提树 U9TB
local lighting 局部采光 U7TA
longitudinal masonry wall 纵墙 U5TA
lower column 下层柱 U5TA

M

main beam 主梁 U5TA
make a living 赚钱过活，谋生 U5TB
make up of 构成，组成 U9TA
make use of 利用 U8TA
masonry structure 砌体结构 U5TA
mathematical proportion 数学比例 U1TA
Memorial Arch (Paifang) 牌坊 U2TA
merge into 汇合，（使）并入；归并 U5TB
Milky Way 银河 U7TB
minor earthquakes 轻微地震 U6TB
mix truck 混凝土搅拌车 U6TA
multi-eaved pagoda 多重檐佛塔 U2TB
multi-storied pagoda 楼阁式塔 U2TB

N

natural gas 天然气 U7TA

O

on a small scale 小规模地 U9TA
on the basis of 以……为基础 U2TA
organic substance 有机物质 U4TA
organic topsoil 有机土层 U6TA
originate from 来自……，源于……起源 U7TB

P

photovoltaic equipment 光伏设备 U10TA
protective measures 防护措施 U6TB

R

rather than 而不是 U2TA
reinforced concrete 钢筋混凝土 U4TA
resistant to 对……有抵抗力 U5TB
ring beam 圈梁 U5TA
rise up 兴起 U1TA
rock-cut 岩石切割 U2TB

S

science and technology 科技 U8TA
seismic load 地震荷载 U6TB
seize upon 利用 U1TA
set out 动身；出发；着手；安排 U10TB
sewage treatment plant 污水处理厂 U7TA
shadow play troupe 皮影剧团 U4TB
Shang Dynasty 商朝 U2TA
shear wall structure 剪力墙结构 U5TA
sick building syndrome 病态建筑综合征，大楼病综合征 U10TA
solar thermal collectors 太阳能集热器 U10TB
spring up 跳起；跃起；迅速成长 U7TB
stand out 突出；坚持；超群；向前跨步 U7TB
stand to 坚持，不放弃 U1TB
stark contrast 鲜明对比 U8TB
storage reservoir 蓄水库 U7TA

sun blinds 太阳薄膜 U10TB
sustainable development 可持续发展 U8TA

T

take ... into account 考虑到 U10TB
take advantage of 利用 U10TA
take on 呈现，具有 U1TA
take place 发生，进行 U1TB
tend to 倾向于 U8TA
transform into 把……转变成…… U5TB
triangular plane 三角形平面 U9TB
triumphal arch 凯旋门 U1TA

U

upper column 上层柱 U5TA

V

vertical component 垂直分量 U6TB
vertical line 垂直线 U8TB

W

wall-framed structure 框架-剪力墙结构 U5TA
water chambers 水箱，蓄水池 U1TB
water supply systems 给水系统 U7TA
Western architectural concepts 西洋建筑理念 U2TB
wheat straw board 麦秸板 U4TB
wrought iron 熟铁，锻铁 U9TA

Appendix IV：建筑专业词汇中英文对照

1．DESIGN BASIS 设计依据
计划建议书 planning proposals
设计任务书 design order
标准规范 standards and codes
条件图 information drawing
设计基础资料 basic data for design
工艺流程图 process flowchart
工程地质资料 engineering geological data
原始资料 original data
设计进度 schedule of design

2．STAGE OF DESIGN 设计阶段
方案 scheme, draft
草图 sketch
会谈纪要 summary of discussion
谈判 negotiation
可行性研究 feasibility study
初步设计 preliminary design
基础设计 basic design
详细设计 detail design
询价图 enquiry drawing
施工图 working drawing, construction drawing
竣工图 as-built drawing

3．CLIMATE CONDITION 气象条件
日照 sunshine
风玫瑰 wind rose
主导风向 prevailing wind direction
最大（平均）风速 maximum (mean) wind velocity
风荷载 wind load
最大（平均）降雨量 maximum (mean) rainfall
雷击及闪电 thunder and lightning
飓风 hurricane
台风 typhoon
旋风 cyclone
降雨强度 rainfall intensity
年降雨量 annual rainfall
湿球温度 wet bulb temperature
干球温度 dry bulb temperature
冰冻期 frost period
冰冻线 frost line
冰冻区 frost zone
室外计算温度 calculating outdoor temperature
采暖地区 region with heating provision
不采暖地区 region without heating provision
绝对大气压 absolute atmospheric pressure
相对湿度 relative humidity

4．GENERAL ROOM NAME 常用房间
办公室 office
服务用房 service room
换班室 shift room
休息室 rest room (break room)

起居室 living room
浴室 bathroom
淋浴间 shower
更衣室 locker room
厕所 lavatory
门厅 lobby
诊室 clinic
工作间 workshop
电气开关室 switchroom
走廊 corridor
档案室 archive
电梯机房 lift motor room
车库 garage
清洁间 cleaning room
会议室（正式）conference room
会议室 meeting room
衣柜间 wardrobe
暖风间 H.V.A.C room
饭店 restaurant
餐厅 canteen, dining room
厨房 kitchen
入口 entrance
接待处 reception area
会计室 accountant room
秘书室 secretary room
电气室 electrical room
控制室 control room
工长室 foreman office
开关柜室 switch gear
前室 antecabinet（Ante.）
生产区 production area
马达控制中心 Mcc
多功能用房 utility room
化验室 laboratory room
经理室 manager room
披屋（阁楼）penthouse
警卫室 guard house

5. ROOFING AND CEILING 屋面及天棚

女儿墙 parapet
雨篷 canopy
屋脊 roof ridge
坡度 slope
坡跨比 pitch
分水线 water-shed
二毡三油 2 layers of felt & 3 coats of bitumastic
附加油毡一层 extra ply of felt
檐口 eave
挑檐 overhanging eave
檐沟 eave gutter
平屋面 flat roof
坡屋面 pitched roof
雨水管 downspout, rain water pipe（R.W.P）
汇水面积 catchment area
泛水 flashing
内排水 interior drainage
外排水 exterior drainage
滴水 drip
屋面排水 roof drainage
找平层 leveling course
卷材屋面 built-up roofing
天棚 ceiling
檩条 purlin
屋面板 roofing board
天花板 ceiling board
防水层 water-proof course
检查孔 inspection hole
人孔 manhole
吊顶 suspended ceiling, false ceiling
檐板（窗帘盒）cornice

6. WALL (CLADDING) 墙体（外墙板）

砖墙 brick wall
砌块墙 block wall
清水砖墙 brick wall without plastering
抹灰墙 rendered wall
石膏板墙 gypsum board, plaster board
空心砖墙 hollow brick wall
承重墙 bearing wall
非承重墙 non-bearing wall
纵墙 longitudinal wall
横墙 transverse wall
外墙 external (exterior) wall
内墙 internal (interior) wall
填充墙 filler wall
防火墙 fire wall
窗间墙 wall between window
空心墙 cavity wall
压顶 coping
圈梁 gird, girt, girth
玻璃隔断 glazed wall
防潮层 damp-proof course (D.P.C)
遮阳板 sunshade
阳台 balcony
伸缩缝 expansion joint
沉降缝 settlement joint
抗震缝 seismic joint
复合夹心板 sandwich board
压型单板 corrugated single steel plate
外墙板 cladding panel
复合板 composite panel
轻质隔断 light-weight partition
牛腿 bracket
砖烟囱 brick chimney
勒脚（基座）plinth

7. FLOOR AND TRENCH 地面及地沟

地坪 grade
地面和楼面 ground and floor
素土夯实 rammed earth
炉渣夯实 tamped cinder
填土 filled earth
回填土夯实 tamped backfill
垫层 bedding course, blinding
面层 covering
结合层 bonding (binding) course
找平层 leveling course
素水泥浆结合层 neat cement binding course
混凝土地面 concrete floor
水泥地面 cement floor
机器磨平混凝土地面 machine trowelled concrete floor
水磨石地面 terrazzo flooring
马赛克地面 mosaic flooring
瓷砖地面 ceramic tile flooring
油地毡地面 linoleum flooring
预制水磨石地面 precast terrazzo flooring
硬木花地面 hard-wood parquet flooring
搁栅 joist
硬木毛地面 hard-wood rough flooring
企口板地面 tongued and grooved flooring
防酸地面 acid-resistant floor
钢筋混凝土楼板 reinforced concrete slab (R.C Slab)
乙烯基地面 vinyl flooring
水磨石嵌条 divider strip for terrazzo
地面做2%坡 floor with 2% slope
集水沟 gully
集水口 gulley
排水沟 drainage trench
沟盖板 trench cover
活动盖板 removable cover plate
集水坑 sump pit

孔翻边 hole up stand
电缆沟 cable trench

8. DOORS/WINDOWS 门窗

木（钢）门 wooden（steel）door
镶板门 panelled door
夹板门 plywood door
铝合金门 aluminum alloy door
卷帘门 roller shutter door
弹簧门 swing door
推拉门 sliding door
平开门 side-hung door
折叠门 folding door
旋转门 revolving door
玻璃门 glazed door
密闭门 air-Tight door
保温门 thermal insulating door
镀锌铁丝网门 galvanized steel wire mesh door
防火门 fire door
（大门上的）小门 wicket
门框 door frame
门扇 door leaf
门洞 door opening
结构开洞 structural opening
单扇门 single door
双扇门 double door
疏散门 emergency door
纱门 screen door
门槛 door sill
门过梁 door lintel
上冒头 top rail
下冒头 bottom rail
门边木 stile
门樘侧料 side jumb
槽口 notch
木窗 wooden window
钢窗 steel window
铝合金窗 aluminum alloy window
百叶窗（通风为主）sun-bind, louver（louver, shutter, blind）
塑钢窗 plastic steel window
空腹钢窗 hollow steel window
固定窗 fixed window
平开窗 side-hung window
推拉窗 sliding window
气窗 transom
上悬窗 top-hung window
中悬窗 center-pivoted window
下悬窗 hopper window
活动百叶窗 adjustable louver
天窗 skylight
老虎窗 dormer window
密封双层玻璃 sealed double glazing
钢筋混凝土过梁 reinforced concrete lintel
钢筋砖过梁 reinforced brick lintel
窗扇 casement sash
窗台 window sill
窗台板 window board
窗中梃 mullion
窗横木 mutin
窗边木 stile
压缝条 cover mould
窗帘盒 curtain box
合页（铰链）hinge（butts）
转轴 pivot
长脚铰链 parliament hinge
闭门器 door closer
地弹簧 floor closer
插销 bolt
门锁 door lock
拉手 pull

链条 chain
门钩 door hanger
碰球 ball latch
窗钩 window catch
暗插销 insert bolt
电动开关器 electric opener
平板玻璃 plate glass
夹丝玻璃 wire glass
透明玻璃 clear glass
毛玻璃（磨砂玻璃）ground glass（frosted glass）
防弹玻璃 bullet-proof glass
石英玻璃 quartz glass
吸热玻璃 heat absorbing glass
磨光玻璃 polished glass
着色玻璃 pigmented glass
玻璃瓦 glass tile
玻璃砖 glass block
有机玻璃 organic glass

9. STAIR/LIFT（ELEVATOR）楼梯、电梯

楼梯 stair
楼梯间 staircase
疏散梯 emergency stair
旋转梯 spiral stair（circular stair）
吊车梯 crane ladder
直爬梯 vertical ladder
踏步 step
扇形踏步 winder（wheel step）
踏步板 tread
档步板 riser
踏步宽度 tread width
防滑条 non-slip insert（strips）
栏杆 railing（balustrade）
平台栏杆 platform railing
吊装孔栏杆 railing around mounting hole
扶手 handrail
梯段高度 height of flight
防护梯笼 protecting cage（safety cage）
平台 landing（platform）
操作平台 operating platform
装卸平台 platform for loading & unloading
楼梯平台 stair landing
客梯 passenger lift
货梯 goods lift
客/货两用梯 passenger/goods lift
液压电梯 hydraulic lift
自动扶梯 escalator
观光电梯 observation elevator
电梯机房 lift mortar room
电梯坑 lift pit
电梯井道 lift shaft

10. BUILDING MATERIAL 建筑材料

10.1 Bricks and Tiles 砖和瓦

红砖 red brick
黏土砖 clay brick
瓷砖 glazed brick（ceramic tile）
防火砖 fire brick
空心砖 hollow brick
面砖 facing brick
地板砖 flooring tile
缸砖 clinkery brick
马赛克 mosaic
陶粒混凝土 ceramsite concrete
琉璃瓦 glazed tile
脊瓦 ridge tile
石棉瓦 asbestos tile（shingle）
波形石棉水泥瓦 corrugated asbestos cement sheet

10.2 Lime, Sand and Stone 灰、砂和石

石膏 gypsum

大理石 marble
汉白玉 white marble
花岗岩 granite
碎石 crushed stone
毛石 rubble
蛭石 vermiculite
珍珠岩 pearlite
水磨石 terrazzo
卵石 cobble
砾石 gravel
粗砂 course sand
中砂 medium sand
细砂 fine sand

10.3 Cement/Mortar/Concrete 水泥、砂浆、混凝土

波特兰水泥（普通硅酸盐水泥）Portland cement
硅酸盐水泥 silicate cement
火山灰水泥 pozzolana cement
白水泥 white cement
水泥砂浆 cement mortar
石灰砂浆 lime mortar
水泥石灰砂浆（混合砂浆）cement-lime mortar
保温砂浆 thermal mortar
防水砂浆 water-proof mortar
耐酸砂浆 acid-resistant mortar
耐碱砂浆 alkaline-resistant mortar
沥青砂浆 bituminous mortar
纸筋灰 paper strip mixed lime mortar
麻刀灰 hemp cut lime mortar
灰缝 mortar joint
素混凝土 plain concrete
钢筋混凝土 reinforced concrete
轻质混凝土 lightweight concrete
细石混凝土 fine aggregate concrete
沥青混凝土 asphalt concrete
泡沫混凝土 foamed concrete
炉渣混凝土 cinder concrete

10.4 Facing/Plastering Materials 饰面及粉刷材料

水刷石 granitic plaster
斩假石 artificial stone
刷浆 lime wash
可赛银 casein
大白浆 white wash
麻刀灰打底 hemp cuts and lime as base
喷大白浆两道 sprayed twice with white wash
分格抹水泥砂浆 cement mortar plaster sectioned
板条抹灰 lath and plaster

10.5 Asphalt（Bitumen）and Asbestos 沥青和石棉

沥青卷材 asphalt felt
沥青填料 asphalt filler
沥青胶泥 asphalt grout
冷底子油 adhesive bitumen primer
沥青玛琋脂 asphaltic mastic
沥青麻丝 bitumastic oakum
石棉板 asbestos sheet
石棉纤维 asbestos fiber

10.6 Timber 木材

裂缝 crack
透裂 split
环裂 shake
干缩 shrinkage
翘曲 warping
原木 log
圆木 round timber
方木 square timber
板材 plank

木条 batten
板条 lath
木板 board
红松 red pine
白松 white pine
落叶松 deciduous pine
云杉 spruce
柏木 cypress
白杨 white poplar
桦木 birch
冷杉 fir
栎木 oak
榆木 willow
榆木 elm
杉木 cedar
柚木 teak
樟木 camphor wood
防腐处理的木材 preservative-treated lumber
胶合板 plywood
三（五）合板 3（5）-plywood
企口板 tongued and grooved board
层夹板 laminated plank
胶合层夹木材 glue-laminated lumber
纤维板 fiber-board
竹子 bamboo

10.7 Metallic Materials 金属材料

黑色金属 ferrous metal
圆钢 steel bar
方钢 square steel
扁钢 steel strap, flat steel
型钢 steel section（shape）
槽钢 channel
角钢 angle steel
等边角钢 equal-leg angle
不等边角钢 unequal-leg angle

工字钢 I-beam
宽翼缘工字钢 wide flange I-beam
丁（之）字钢 T-bar（Z-bar）
冷弯薄壁型钢 light gauge cold-formed steel shape
热轧 hot-rolled
冷轧 cold-rolled
冷拉 cold-drawn
冷压 cold-pressed
合金钢 alloy steel
钛合金 titanium alloy
不锈钢 stainless steel
竹节钢筋 corrugated steel bar
变形钢筋 deformed bar
光圆钢筋 plain round bar
钢板 steel plate
薄钢板 thin steel plate
低碳钢 low carbon steel
冷弯 cold bending
钢管 steel pipe（tube）
无缝钢管 seamless steel pipe
焊接钢管 welded steel pipe
黑铁管 iron pipe
镀锌钢管 galvanized steel pipe
铸铁 cast iron
生铁 pig iron
熟铁 wrought iron
镀锌铁皮 galvanized steel sheet
镀锌铁丝 galvanized steel wire
钢丝网 steel wire mesh
多孔金属网 expanded metal
锰钢 managanese steel
高强度合金钢 high strength alloy steel

10.8 Non-Ferrous Metal 有色金属

金 gold
白金 platinum

铜 copper
黄铜 brass
青铜 bronze
银 silver
铝 aluminum
铅 lead

10.9 Anti-Corrosion Materials 防腐蚀材料

聚乙烯 polythene, polyethylene
尼龙 nylon
聚氯乙烯 PVC（polyvinyl chloride）
聚碳酸酯 polycarbonate
聚苯乙烯 polystyrene
丙烯酸树脂 acrylic resin
乙烯基酯 vinyl ester
橡胶内衬 rubber lining
氯丁橡胶 neoprene
沥青漆 bitumen paint
环氧树脂漆 epoxy resin paint
氧化锌底漆 zinc oxide primer
防锈漆 anti-rust paint
耐酸漆 acid-resistant paint
耐碱漆 alkali-resistant paint
水玻璃 sodium silicate
树脂砂浆 resin-bonded mortar
环氧树脂 epoxy resin

10.10 Building Hardware 建筑五金

钉子 nails
螺纹屋面钉 spiral-threaded roofing nail
环纹石膏板钉 annular-ring gypsum board nail
螺丝 screws
平头螺丝 flat-head screw
螺栓 bolt
普通螺栓 commercial bolt
高强螺栓 high strength bolt
预埋螺栓 insert bolt
胀锚螺栓 cinch bolt

垫片 washer

10.11 Paint 油漆

底漆 primer
防锈底漆 rust-inhibitive primer
防腐漆 anti-corrosion paint
调和漆 mixed paint
无光漆 flat paint
透明漆 varnish
银粉漆 aluminum paint
磁漆 enamel paint
干性油 drying oil
稀释剂 thinner
焦油 tar
沥青漆 asphalt paint
桐油 tung oil, Chinese wood oil
红丹 red lead
铅油 lead oil
腻子 putty

11．OTHER 其他建筑术语

11.1 Discipline 专业

建筑 architecture
土木 civil
给排水 water supply and drainage
总图 plot plan
采暖通风 H.V.A.C
电力供应 electric power supply
电气照明 electric lighting
电讯 telecommunication
仪表 instrument
热力供应 heat power supply
动力 mechanical power
工艺 process technology
管道 piping

11.2 Conventional Terms 一般通用名词

建筑原理 architectonics
建筑形式 architectural style

民用建筑 civil architecture
城市建筑 urban architecture
农村建筑 rural architecture
农业建筑 farm building
工业建筑 industrial building
重工业的 heavy industrial
轻工业的 light industrial
古代建筑 ancient architecture
现代建筑 modern architecture
标准化建筑 standardized buildings
附属建筑 auxiliary buildings
城市规划 city planning
厂区内 within site
厂区外 offsite
封闭式 closed type
开敞式 open type
半开敞式 semi-open type
模数制 modular system
单位造价 unit cost
概算 preliminary estimate
承包商 constructor, contractor
现场 site
扩建 extension
改建 reconstruction
防火 fire-prevention
防震 aseismatic, quake-proof
防腐 anti-corrosion
防潮 dump-proof
防水 water-proof
防尘 dust-proof
防锈 rust-proof
车流量 traffic volume
货流量 freight traffic volume
人流量 pedestrian volume
透视图 perspective drawing
建筑模型 building model

11.3 Architectural Physics 建筑物理

照明 illumination
照度 degree of illumination
亮度 brightness
日照 sunshine
天然采光 natural lighting
光强 light intensity
侧光 side light
顶光 top light
眩光 glaze
方位角 azimuth
辐射 radiation
对流 convection
传导 conduction
遮阳 sun-shade
保温 thermal insulation
恒温 constant temperature
恒湿 constant humidity
噪声 noise
隔音 sound-proof
吸音 sound absorption
露点 dew point
隔汽 vapor-proof

11.4 Name of Professional Role 职务名称

项目经理 project manager （PM）
设计经理 design manager
首席建筑师 principal architect
总工程师 chief engineer
土木工程师 civil engineer
工艺工程师 process engineer
电气工程师 electrical engineer
机械工程师 mechanical engineer
计划工程师 planning engineer
助理工程师 assistant engineer
实习生 probationer
专家 specialist, expert

制图员 draftsman
技术员 technician

11.5 Drafting 制图

总说明 general specification
工程说明 project specification
采用标准规范目录 list of standards and specification adopted
图纸目录 list of drawings
平面图 plan
局部放大图 detail with enlarged scale
……平面示意图 schematic plan of...
……平剖面图 sectional plan of...
留孔平面图 plan of provision of holes
剖面 section
纵剖面 longitudinal section
横剖面 cross（transverse）section
立面 elevation
正立面 front elevation
透视图 perspective drawing
侧立面 side elevation
背立面 back elevation
详图 detail drawings
典型节点 typical detail
节点号 detail No.
首页 front page
图纸目录及说明 list of contents and description
图例 legend
示意图 diagram
草图 sketch
荷载简图 load diagram
流程示意图 flow diagram
标准图 standard drawing
……布置图 layout of ...
地形图 topographical map
土方工程图 earth-work drawing
展开图 developed drawing
模板图 formwork drawing
配筋 arrangement of reinforcement
表格 tables
工程进度表 working schedule
技术经济指标 technical and economical index
建、构筑物一览表 list of buildings and structures
编号 coding
序列号 serial No.
行和栏 rows and columns
备注 remarks
等级 grade
直线 straight line
曲线 curves
曲折线 zigzag line
虚线 dotted line
实线 solid line
影线 hatching line
点划线 dot and dash line
轴线 axis
等高线 contour line
中心线 center line
双曲线 hyperbola
抛物线 parabola
切线 tangent line
尺寸线 dimension line
圆形 round
环形 annular
方形 square
矩形 rectangle
平行四边形 parallelogram
三角形 triangle
五角形 pentagon
六角形 hexagon

八角形 octagon
梯形 trapezoid
圆圈 circle
弓形 sagment
扇形 sector
球形的 spherical
抛物面 paraboloid
圆锥形 cone
椭圆形 ellipse, oblong
面积 area
体积 volume
容量 capacity
重量 weight
质量 mass
力 force
米 meter
厘米 centimeter
毫米 millimeter
公顷 hectare
牛顿/平方米 Newton/square meter
千克/立方米 kilogram/cubic meter
英尺 foot
英寸 inch
磅 pound
吨 ton
加仑 gallon
千磅 kip
平均尺寸 average dimension
变尺寸 variable dimension
外形尺寸 overall dimension
展开尺寸 developed dimension
内径 inside diameter
外径 outside diameter
净重 net weight
毛重 gross weight

数量 quantity
百分比 percentage
净空 clearance
净高 headroom
净距 clear distance
净跨 clear span
截面尺寸 sectional dimension
开间 bay
进深 depth
单跨 single span
双跨 double span
多跨 multi-span
标高 elevation, level
绝对标高 absolute elevation
设计标高 designed elevation
室外地面标高 ground elevation
室内地面标高 floor elevation
柱网 column grid
坐标 coordinate
厂区占地 site area
使用面积 usable area
辅助面积 service area
通道面积 passage area
管架 pipe rack
管廊 pipeline gallery
架空管线 overhead pipeline
排水沟 drain ditch
集水坑 sump pit
喷泉 fountain
地漏 floor drain
消火栓 fire hydrant
灭火器 fire extinguisher
二氧化碳灭火器 carbon dioxide extinguisher
卤代烷灭火器 halon extinguisher

Appendix V：建设工程施工合同

发包方（甲方）：××××
Party A：××××

承包方（乙方）：××××
Party B：××××

本合同由如上列明的甲、乙双方按照《中华人民共和国合同法》《建筑安装工程承包合同条例》以及国家相关法律法规的规定，结合本合同具体情况，于××年×月×日在××签订。

This contract is signed by the two Parties in ×××× on ×××× according to the "Contract Law of the People's Republic of China", the "Regulation on Building and Installation Contracting Contract", and other relevant national laws and regulations, as well as the specific nature of this project

第一条 总则
Article 1 General Principles

1.1 合同文件
Contract Documents

本合同包含合同、展示、施工图纸、作法说明、招投标文件以及合同所指的其他文件，它们均被视为合同的一部分。合同文件应该能够相互解释，互为说明。组成本合同的文件及优先解释的顺序如下：本合同、中标通知书、投标书及其附件、标准、规范及有关技术文件、图纸、工程量清单及工程报价单或预算书。合同履行中，发包人承包人有关工程的洽商、变更等书面协议或文件视为本合同的组成部分。

This contract includes the contract itself, exhibits, construction drawings, explanations of work procedures, tender documents, as well as other documents specified by the contract. The contract documents should be mutually explainable and mutually descriptive. The documents that constituted the contract and the priority hierarchy of explanations of the documents are as follows: this contract; the official letter of winning the tender; Letter of bidding and its annex, standards, guidelines, and relevant technical documents, drawings, Bill of Quantity (BOQ), quotation of prices, and project budget. During the implementation of the contract, any written agreements or documents between two parties such as change-orders will be deemed as part of this contract.

1.1.1 项目概况

Project Introduction

1.1.2 工程名称：××××

Name of Project: ××××

1.1.3 工程地点：××××

Location of Project: ××××

1.1.3.1 工程范围：除了本合同其他条款另有说明外，乙方应提供为履行合同所需的所有服务、用品以及其他保障工程进行所需的必要花费以完成下列工作：

Scope of the project: unless otherwise explained by other articles in this contract, Party B shall provide all services, utensils, and other necessary cost for the implementation of this contract; the scope of project includes, but not limited to:

※室内装饰工程，详见附件一和附件二

　　Inner Decoration, see appendix 1 & 2

※室外装饰工程（门窗、屋面、台阶等），详见附件一和附件二

　　Facade(including doors, windows, roof and stairs) see appendix 1 & 2

※电器工程，详见附件一和附件二

　　Electronic Engineering, see appendix 1 & 2

※空调工程，详见附件一和附件二

　　Air Conditioning System, see appendix 1 & 2

※给排水及采暖工程，详见附件一和附件二

　　Water Supply, Sewage and Heating System, see appendix 1 & 2

※消防工程

　　Fire Control System, see appendix 1 & 2

※其他图纸上所列项目（任何对原定设计的变更应提供新的图纸并有甲方的签字确认）以上所有的工程项目以下统称为"工程"。

　　Other items listed in the construction drawings (any changed drawings shall be provided and confirmed by Party A). All these aforementioned items are generally referred as "project" hereafter.

1.1.3.2 承包方式：乙方以包工包料，包工程质量，包安全，包文明施工的方式承包工程

Nature of Contracting: Party B shall be responsible for both labor and materials for this project, and will responsible for the quality of this project, safety, as well as code of conduct during the construction project.

1.1.4 甲方工地代表由甲方指派

On-site representative of Party A shall be appointed by Party A

1.1.5 开工日期××××

Date of the project beginning: ××××

1.1.6 竣工日期××××

Date of the project completion: ××××

1.1.7 工程质量：质量合格，达到本合同要求和国家质量验收标准，并确保竣工验收一次合格。验收内容包括但不仅限于：①隐蔽工程；②分项工程；③实验报告；④材料检验报告。如果由于验收未能一次通过而导致后续工程开工的延迟，乙方将按照每日工程总款的2‰金额赔偿给甲方作为罚金，但最多不超过工程总价款的5%。

Quality of Project: The quality should be "Excellent", and shall meet the requirements by this contract and the National Quality Standards, and should be guaranteed to be accepted at the first attempt upon project completion, and shall also LEED certified. The scope of as-built acceptance includes, but not limit to: ①concealed work; ②itemized project; ③reports of experiments; ④reports of material testing. In the event of project fails to be accepted at the first inspection and thus incurs delays beginning of the subsequent projects. Party B shall compensate Party A at the daily rate of 2‰ of the total contract amount as penalty, but the total amount shall not exceed 5% of the total contract amount.

1.1.8 工程资质：在实施此合同所涉及的工程时，乙方应确保其具有相关资质。如乙方因为不具备相关资质导致甲方的损失或工程进度的滞后，乙方承担全部责任。

Qualifications: Party B represents and warrants it is competent and legitimate to carry out the services that set forth under this Agreement when implementing this contract. Party B shall be responsible for any delay of the construction induced by their insufficient qualifications.

1.2 合同价款

Contract Amount

本合同价款为人民币××××，为固定总价合同价款。

The total contract amount is ××××, which is fixed contract amount.

本合同包括设备材料费、人工费、机械费、管理费、利润、税金、代办费用、保险费、运输费、劳务费用、总承包服务费以及为完成工程所必需的其他一切费用，还有工程各系统检验检测、安装调试等费用。该款项为甲方应付给乙方的全部费用，并已经包含物价变动因素。

This agreement includes the cost of equipment and materials, labor cost, machinery cost, management cost, profit, tax, government charges on behalf of Party A, insurance, transportation, labor, subcontract service fee, and all other cost for the completion of the project, as well as every system testing, installation, and commissioning costs. This total contract amount is exact amount that Party A shall pay to Party B and it includes the fact of price fluctuation.

1.3 工程结算

Final Audit of project

1.3.1 本工程计价依据GB50500-2013《建设工程工程量清单计价规范》采用固定总价，除施工期间甲方要求变更现有图纸及发生的增减项洽商，甲方将作价格调整。

This project is using fixed total amount according to GB50500—2013 the Pricing Guidelines by Code of Quantity in Construction Projects, except that Party A may make price adjustment in the event of the existing construction drawings are changed or add or remove any construction work according to the request of Party A.

1.3.2 所有的人工费用都不得以任何理由进行过调整，结算时只接受投标文件中约定的价格。

All labor cost shall NOT be adjusted in any event. Only the prices stimulated in the bidding documents are accepted at the final audit after completion.

1.3.3 因工程调整而产生的工程洽商，如洽商费用超过人民币五万元整，则洽商文件必须在工程中期验收当日被提交给甲方，所产生的洽商费用在交付中期款时一并核算结清。如洽商费用低于五万元人民币整，则洽商文件必须在工程验收当日内提交给甲方，所产生的费用在交付工程验收款时一并核算结清。

If the cost of change orders are over RMB 50,000 Yuan, then all change orders will be issued because the construction adjustments shall be submitted to Party A on the same day of the construction mid-term check and acceptance, and the approved amount will be paid to Party B with the construction mid-term payment; if the costs of the change orders are less than 50,000 Yuan, then all change orders will be issued because the construction adjustments shall be submitted to Party A on the same day of the construction completion check and acceptance, and the approved amount will be paid to Party B with the construction completion payment.

第二条 政府批复

Article 2 Government Approvals

2.1 乙方负责协助甲方办理此项施工需要办理的所有政府报批，包括但不仅限于消防、环保的报批与其他与本工程相关的报批，甲方负责政府行政审批的费用。

Party B shall be responsible to process government approvals related to the project, including but not limited to: fire control, environment protection and other approvals related to this Project. Party A is responsible for the costs of these approvals.

2.2 乙方负责其工程报备

Party B shall be responsible for construction applications, approvals and related costs.

2.3 乙方应承担2.2所述报备不全而导致的政府罚金，并且不能顺延工期。

Party B shall be responsible for any government fines due to incomplete approval processes, and should NOT cause delay in the construction period.

第三条 甲方的权利和义务

Article 3 Rights and Obligations of Party A

3.1 向乙方提供经确认的施工图纸或做法说明五份，并向乙方进行现场交底，向乙方提供施工所需要的水电设备，并说明使用注意事项。

Party A shall provide FIVE copies of confirmed construction drawings and explanations of work procedures, and shall conduct technical clarification. Party A shall provide Party B with necessary water and power equipment and the instructions on using the equipment.

3.2 指派××××为甲方驻工地代表，负责合同的履行。对工程质量、进度进行监督检查，办理验收、变更事宜。甲方的工地代表如果认为乙方的某雇员行为不当或散漫不羁，或甲方认为该雇员不能胜任，则乙方应立即进行撤换，并且未经甲方实现书面许可，乙方不得在施工中再行雇佣该等人员。被撤换的任何人员应尽可能立即由甲方代表所批准的胜任的人员代替。

Party A appoints ×××× as the on-site representative to supervise the fulfillment of contract obligations, to oversee the quality control and check the project progress, and to handle issues such as acceptance and variations. In the event of the on-site representative of Party A believes that any employee of Party B has improper conduct or misbehaviors, or if Party A consider such employee is not competent for the relevant work, Party B shall dismiss and replace this employee immediately; and unless having received consent by Party A in writing in advance, Party B shall no longer hire the specified employee in the following phase of this project. Party B should make every effort to replace the dismissed employee with competent personnel authorized by on-site representative of Party A.

3.3 甲方按照付款约定向乙方支付工程款。

Party A should make payments to Party B according to the payment terms.

3.4 如甲方认为乙方确已无能力继续履行合同的，则甲方有权解除合同，乙方必须在接到甲方书面通知后两周内撤离场地以便乙方继续施工。对乙方完成工程量的结算不作为撤离场地的条件，结算应在竣工一个月之内完成，付款方式按照本合同条款执行。

In the event of Party A consider Party B is indeed unable to further fulfill its contract obligations, then Party A has the right to terminate the contract, and Party B should vacate the project site within two weeks after receiving Party A's notice, so that Party A can resume construction as soon as possible. Confirmation of finished quantity shall NOT be regarded as a prerequisite to Party B's vacation of the project site, and the confirmation should be processed within one month after the project completion. Payment shall be made according to the payment terms in this contract.

3.5 甲方有权要求乙方按照甲方的组织设计和施工工期进行施工,乙方如对甲方的要求和指令不予执行的,则按照乙方违约处理,乙方应承担违约责任。

Party A has the right to instruct Party B to do the construction according to Party A's organization and planned construction period; in the event of Party B ignores and refuses to observe Party A's instructions, Party B will be deemed as breach of the contract, and Party B shall be held responsible for the liabilities of breaching the contract.

3.6 甲方有权要求乙方服从甲方的统一安排和管理,乙方的施工人员必须遵守各项规章制度。对于乙方施工人员违规操作所产生的后果,乙方承担全部责任。甲方有权从工程总价款中扣除相应的款项。

Party A has the right to instruct Party B to observe the general arrangement and management by Party A, and the construction staffs of Party B must observe all rules and regulations. Party B shall be fully responsible for the consequences of such violations by the staff of Party B. Party A has the right to directly deduct the corresponding amount from the total contract amount.

3.7 甲方所有的通知可以以书面或口头形式发出,乙方接获甲方的通知后应根据通知规定的内容执行。

All notices from Party A can be made both in writing and orally, but the oral instructions are only applicable to the representative of Party A. Party B shall, upon receiving notices from Party A, execute the corresponding matters according to the content of the notices.

3.8 未经甲方同意,乙方不得将工程转让或分包,对于乙方将工程擅自转让或分包的,甲方有权解除合同,乙方必须接受并承担由此造成的一切后果。

Party B shall NOT transfer or subcontract the project to any third party without the consent of Party A; in the event of Party B transfers or subcontracts the project to any third party without the consent of Party A, Party A has the right to terminate the contract, and Party B must accept and be responsible for any consequence of such violation.

3.9 甲方有权指定专业分包商,并按照乙方在投标文件中承诺的配合管理费率向乙方支付配合管理费。

Party A has the right to appoint special sub-contractors, and shall pay the management fee to Party B according to the rate of subcontract management and coordination fee stipulated in the bidding documents.

3.10 按照招标文件中的相关条目,甲方有权自行采购材料供给乙方,并按照乙方在投标文件的相关条目中汲取人工费、机械费、辅助材料费等向乙方支付费用。甲方没有义务向乙方支付其他费用。

Party A has the right to purchase and provide Party B with materials on its own, and shall pay labor cost, machinery cost, and accessory cost to Party B according to the relevant terms in the bidding documents. Party A has no obligations to pay any other costs to Party B.

第四条 乙方的权利义务
Article 4 Rights and Obligations of Party B

4.1 参加甲方组织的施工图纸或做法说明的现场交底，拟定施工方案和进度计划，交由甲方审定。施工过程中，乙方按照甲方的有关指示及修改变更要求进行相应调整。

Party B shall participate in the technical clarification on construction drawings and explanations of work procedures organized by Party A, and should develop the construction plan and the progress schedule for Party A to review. During construction, Party B should make such adjustment to the construction plan and the progress schedule per Party A's instruction and change-orders.

4.2 指派××××为乙方驻工地代表，负责合同的履行，按照要求组织施工，保质保量、按期完成施工任务。解决由乙方负责的各项事宜。乙方驻工地代表在本公司的工作时间不得少于90%工作日。

Party B appoints ×××× as the on-site representative of Party B to take charge of fulfilling the contract obligations. The on-site representative of Party B shall organize the construction according to contract requirements so that the project can be completed according to the quality and quantity requirements and on a timely manner. The on-site representative of Party B shall work on the project site for no fewer than 90% of the total working days.

4.3 严格执行施工规范、安全操作规程、防火安全规定、环境保护规定。严格按照图纸或做法说明进行施工，做好各项质量检查记录，参加竣工验收，编制竣工资料及工程结算。

Party B shall strictly observe construction guidelines, safety requirements, fire safety regulations, environment protection regulations and LEED requirements. Party B shall do the construction by strictly adhering to the construction drawings and explanations of work procedures, and shall effectively record each QA/QC inspection. Party B shall attend the completion acceptance, and develop at-built submittals and final project audit.

4.4 根据甲方和设计方一致通过的设计图纸进行施工，遵从甲方工地代表和设计方的指示，如乙方发现在工程标准和设计图之间存在不一致时，应以书面形式通知甲方和设计方。

Party B shall carry out the construction according to the design drawings agreed upon by both Party A and the designer, shall observe the instructions of designer and the representative of Party A; in the event of Party B finds inconsistence between project standards and design drawings,

Party B should inform the designer and Party A in writing.

4.5 遵守国家或地方政府及有关部门对施工现场管理的规定及甲方的规章制度，妥善保护好施工现场周围的建筑物、设备管线等不受损坏。做好施工现场保卫、消防和垃圾消纳等工作，处理好由于施工带来的扰民问题及周围单位（住户）的关系；如果由于乙方导致任何设施、设备或管道在工地内外受损，将由乙方赔偿损失。乙方负责与甲方的施工协调事项，并独自完全承担所有相关责任。

Party B shall observe policies of on-site management by the state and local government authorities as well as Party A's rules and regulations; Party B shall compensate for the losses to any facilities, equipment or pipelines both inside and outside of the construction yard caused by party B. Party B shall be responsible for the coordination of construction issues with party A, and shall independently and fully assumes corresponding responsibilities.

4.6 任何原建筑物结构的拆改或设备管道的挪移，乙方应向甲方申报；在甲方批复前，不得擅自动工。

Party B should report to party A of any demolition of or variation to any original building structure or re-position of equipment pipes; Party B will refrain from any such work prior to receiving Party A's approval

4.7 乙方应保护所有设备以及工程成品，并承担竣工移交前所引起损坏的损失赔偿。

Party B should protect all equipment and finished parts, and shall assume compensation responsibilities of any losses prior to completion hand-over.

4.8 乙方负责用扩展保险责任批单为与建筑物相关的工程以及工程所需的材料购买工程一切险，乙方及乙方分包商（以下简称"分包商"）应对工程进行中自己或其他方的财产出资购买相应的保险，这些财产包括：工棚、铁架塔、脚手架、工具及其他的一些材料。

Party B shall purchase Construction All Risks Insurance for the project and required materials related to the building structure by using extended coverage endorsement; during construction, Party B and its subcontractors (hereafter abbreviated as "subcontractors") should purchase corresponding insurance for their own or other parties' properties, such as temporary construction barracks, iron frames and towers, scaffolding, utensils and other materials.

4.9 从工程开始至工程竣工验收报告上所列明的日期之日止，乙方应全权负责整项工程。不管任何原因导致的工程的损坏，乙方负责修理还原，以使整个工程在竣工时能完全依照合同要求以及甲方工地代表的指示，状况良好。

From the beginning of construction to the completion date stipulated by the as-built acceptance report, Party B will be fully responsible for the entire project. Party B shall fix and reinstate and damages to the project items regardless of the cause, so that the project shall be in good condition upon completion, according to the contract requirements and the instructions of Party A's representative.

4.10 除非合同另有规定，乙方应就下列事项独自完全承担如下赔偿责任，并确保甲方完全免责：与因工程施工及维修而对人身造成的任何伤害或损害或对财产造成任何损害有关的所有损失及权利要求。但在下列情形下的赔偿或补偿除外：

Unless otherwise specified, Party B should independently and fully assume the following compensation liabilities, and exempt Party A from all liabilities: all injury and rights claims related to any personal injuries or property damage or loss as a result of the construction project, except for the following compensation or reimbursement:

4.10.1 工程或其任何部分永久性使用或占用工地。

Permanent use or occupation of land by the project or any part of the project.

4.10.2 甲方享有在任何土地上、下、里穿越其施工的权利。

Party A has the right to traverse the construction site from above, beneath, or inside the site.

4.10.3 根据合同规定对财产的损伤或伤害是工程施工或维修之不可避免的后果。

The property damage or loss is an inevitable loss caused by the construction or repair according to the contra-requirements.

4.11 乙方保证施工现场清洁，符合环境卫生管理的有关规定，交工前清理现场达到甲方要求，如达不到甲方的要求，乙方承担因自身违反规定造成的损失和罚款。

Party B shall keep the construction site clean and meet the regulations of environment sanitation, and make sure the condition of the site meets the requirements of Party A. If the condition of the construction site does NOT meet the requirements of Party A, Party B shall be responsible for any loss or fine as a result of its own violation of such regulations.

4.12 乙方必须要按照施工图所显示的设计要求施工，并符合施工规范规定。所有因对施工图的误解或疏忽引起的施工错误，须返工者，其一切费用均由乙方自负，且不得因此延长完工期限。

Party B MUST adhere to the design requirements in the construction drawings, and must meet the requirements of construction guidelines. Party B shall be fully responsible for any cost of rework on construction mistakes as a result of misunderstanding of the construction drawings or negligence, and shall NOT delay the completion of the project.

4.13 乙方应与每位施工人员订立务工合同，办理合法务工证件并承担费用。

Party B should sign labor contract with all its construction staffs, and issue proper work. Be responsible for the corresponding cost.

4.14 乙方应每月按时向每位施工人员发放工资，如果乙方未按规定发放工资而给甲方带来任何损失，则乙方承担完全赔偿责任。

Party B should make salary/wages payment to all construction staffs in time on a monthly basis, and will be fully responsible for any liabilities caused by any loss or damage to Party A as a result of Party B's failure to pay salaries and wages according to the requirements.

4.15 在履行本合同过程中如对任何第三人造成任何侵害和损失，或违反任何相关国家法律法规，乙方对其行为和行为后果承担全部责任。

Party B shall take full responsibility for its actions or consequences caused by its actions which cause infractions to any third party and cause any violations to related state laws and regulations when implementing this contract.

4.16 乙方应对本工程所有的分包商负有管理和提供总包服务的义务,包括但不限于材料报审、工程预验收和报审、分包付款申请的审批、工程资料编制等。

Party B shall take the responsibilities of sub-contractor management and providing services designated to general contractor, which includes but not limited to: submitting and acquiring construction materials' approval, construction pre-examination and submitting for approval, review and approve payment applications from sub-contractors and construction document filing, etc.

第五条 付款条约
Article 5 Payment Terms

5.1 双方商定本合同价款编制采用工程量清单计划规范,固定合同总价,除施工期间甲方要求变更现有图纸,甲方将作价格调整。

The parties agree to use fixed total amount according to GB50500—2013 the Pricing Guidelines by Code of Quantity in Construction Projects, except that Party A may make price adjustment in the event that the existing construction drawings are changed according to the request of Party A.

5.2 本工程付款方式:工程开工后5个工作日内甲方支付总工程款的30%,室外门窗及幕墙工程验收合格后5个工作日内甲方支付总工程款的20%,室内中间验收完成后5个工作日内甲方支付总工程款的20%,工程完工验收合格乙方向甲方提交完整的竣工资料后5个工作日内支付工程总价款的27%,工程完工验收合格保修期一年后5个工作日内支付工程总价款的70 000元。

Payment terms: Party A shall pay 30% of the total contract amount to Party B within 5 working days after signing of the contract; 20% of the total contract amount shall be paid within 5 working days after the facade is finished; 20% of the total contract amount shall be paid within 5 working days after the construction mid-term check and acceptance is completed; 27% of the total contract amount 70, 000 Yuan shall be paid within 5 workings after the construction completion check and acceptance.

第六条 开始与延期
Article 6 Beginning and Delaying of Project

6.1 整个工程应在开工后××××天内完成。

The whole project should be completed within ×××× calendar days after the beginning of the project.

6.2 非本合同规定的可以延期的情形外，如果甲方代表认为工程或其部分工程的进度太过缓慢以致无法保证工程在规定时间或延长的完工期间内完工，乙方应随即采取必要的措施，甲方代表可批准加快进度以在规定时间或约定的延长期内完成工程或其任何部分。乙方无权因采取该措施而获得额外支付。

Unless otherwise specified by the contract as acceptable conditions of delay, in the event of the on-site representative of Party A believes that the progress of the project or any part of the project substantially lags behind schedule, which may cause Party B unable to complete the project within the required construction period or any extended construction period agreed upon, Party B shall promptly take necessary measures under the authorization of the on-site representative of Party A to expedite the progress so that the project or any part of the project could be completed within the required construction period or any extended construction period agreed upon. Party B is NOT subject to any additional compensation for the measures taken.

6.3 如果由于非乙方原因导致的工程停工，比如工地停电、停水，该情形持续每满8小时，竣工日期往后顺延一天，乙方应有书面记录并经甲方代表签字。

Suspension of the construction work as a result of non-Party B's factors, such as power failure and water supply cut-off, the completion date shall be postponed by one day for every 8 hours of such conditions. Party B should inform Party A immediately in such events and make record of the condition in writing for the representative of Party A to sign off.

第七条　保修

Article 7 Warranty

7.1 乙方保证以一流精湛的工艺完成工程，并保证除其他机械或设备外，整个工程无瑕疵，但隐性瑕疵不受此限。乙方应按相关规定对该工程在设计使用年限内提供保修。

Party B shall complete the project with excellent workmanship and guarantees that, except for other machines or equipment, the project as a whole will be free of defect within eighteen months after the completion acceptance or prior to the last payment (whichever is later); warranty on hidden defects is not limited to this date. The warranty period shall be in accordance with the relative regulations and designed working life.

7.2 在工程建设过程中，或上述保修期内，乙方应及时修复材料或人工的瑕疵，由此引发的损失或损坏将由乙方完全赔付。

During the whole construction period or the aforementioned warranty period, Party B shall repair any defects of material and workmanship on a timely manner; and Party B shall be fully responsible for any losses or damage caused by such defects.

7.3 在保修期内，乙方指派专人负责维护。隐性的瑕疵应在发现问题后24小时内修复，影响甲方正常运营的应在4小时内到现场修复。如果乙方未能及时修复，甲方有权利自行决定雇佣第三方进行修复，由此发生的费用将由乙方来承担。

During the warranty period, Party B shall appoint special personnel for the maintenance. Hidden defects should be fixed within 24 hours after being detected, and those that have an influence on Party A's normal operation should be fixed within 4 hours on the spot. In the event of Party B fails to fix the defects on a timely manner, Party A has the right to hire a third party to do the repair, and any cost incurred shall be burdened by Party B.

第八条 关于工程质量及验收的约定
Article 8 Agreement on Quality and Acceptance

8.1 双方同意本工程将以施工图纸、作法说明、设计变更和《建筑装饰和装修工程质量验收规范》（GB 50210—2018）、《建筑工程质量验收统一标准》（GB 50300—2013）、《建筑电气工程施工质量验收规范》（GB 50303—2015）、《通风与空调工程施工质量验收规范》（GB 50243—2016）、《建筑给排水及采暖工程施工质量验收规范》（GB 50242—2016）等国家制定的施工及验收规范为质量验收标准。

The parties agrees that the construction drawings, explanations of work procedures, design change-orders and such national guidelines on construction and acceptance as the Quality Acceptance Criteria of Construction, Decoration, and Remodeling Projects (GB 50210—2018), the General Quality Acceptance Standards of Construction Projects (GB 50300—2013), the Quality Acceptance Guidelines of Mechanics and Engineering Projects of Construction (GB 50303—2015), the Quality Acceptance Guidelines of HVAC Projects (GB 50243—2016), the Quality Acceptance Guidelines of Water Supply, Sewage, and Heating Systems in Construction (GB 50242—2016) are regarded as the acceptance standards for the project.

8.2 本工程质量应达到国家质量验收标准，并确保一次验收合格。

This project should meet the national quality acceptance standards and should be accepted at the first inspection after completion.

8.3 甲、乙双方应合力办理隐蔽工作和中间工程的检查与验收手续。乙方应提前24小时书面通知甲方参加各类验收，甲方如果未能及时参加隐蔽工程和中间工程验收，乙方可自行验收，甲方应予承认。若甲方要求复验时。乙方应按照要求办理复验。若复验合格，甲方应承担复验费用，由此造成停工的，工期顺延；若复验不合格，其复验费用由乙方承担，且工期不予顺延。

Two Parties shall jointly undertake the mid-term inspection and acceptance process for concealed work. Party B shall inform Party A in writing of all kinds of inspection and acceptance

24 hours in advance. In the event of Party A fails to participate in the mid-term acceptance of concealed work, Party B can do the acceptance independently; Party A should accept such acceptance. When Party A requests a re-inspection, Party B shall carry out the re-inspection accordingly. If the result of re-inspection has passed acceptance standards, Party A shall be responsible for the re-inspection cost, and the project period shall be extended by the duration of project suspension caused by such re-inspections; If the result of re-inspection has NOT passed acceptance standards, Party B shall be responsible for the re-inspection cost, and the project period shall NOT be extended.

8.4 乙方应确保甲方对于因乙方所提供的劣质材料而引发的任何事件后果完全免责，其返工费用由乙方承担，工期不予顺延。

Party B should guarantee that Party A is fully exempted from liabilities for any events as a result of materials with poor quality; the cost of re-work shall be burdened by Party B, and the project period shall NOT be extended.

8.5 工程竣工后，乙方应通知甲方验收，甲方自接到验收通知3个工作日内组织验收，并办理验收、移交手续。如甲方在规定时间内未能组织验收，需及时通知乙方，另行确定验收日期。

Upon completion of project, Party B should inform Party A to process completion acceptance; Party A shall arrange inspection and acceptance within 3 working days after receiving the notification, and shall handle the inspection and hand-over processes. In the event of Party A could NOT arrange the inspection and acceptance within the required period, Party A should inform Party B and reschedule the date for acceptance on a timely manner.

8.6 乙方在向甲方提交竣工验收申请的同时，应提交相关的竣工资料和竣工图。

Party B shall submit relevant as-built submittals and as-built drawing when submitting the request for completion acceptance to Party A.

8.7 竣工验收后，乙方必须在7天内，完成场地清退工作，并向甲方做好各项交接工作包括文件、设备和工具等。

After the completion acceptance, Party B must finish the vacation of the project site within 7 days and complete all hand-over processes including documents, equipment, and utensils.

8.8 甲方代表可对每一工作日的工程质量进行检查检验。甲方代表对工程质量提出整改或返工通知后，乙方必须在规定时间内按要求完成。

The on-site representative of Party A may carry out the inspection of the project quality on any single work day. Having received the notification of correction or re-work due to inadequate quality from the representative of Party A, Party B shall complete the work within the required time frame.

第九条 关于材料设备的约定

Article 9 Agreement on the Supply of Materials and Equipment

9.1 本工程甲方负责采购供应的材料、设备（见附表），应为全新的合格产品，并应按时供应到现场。凡约定由乙方提货的，甲方应将提货手续移交给乙方，由乙方承担运输费用。如果由于甲方提供的材料设备质量低劣或规格差异，对工程造成损失，责任由甲方承担。甲方供应的材料，经乙方验收后，由乙方负责保管。由于乙方保管不当或被盗所造成的损失，由乙方负责赔偿。

All materials and equipment (see the attached table) that Party A orders should be brand new and qualified products, and should be delivered to the construction site as scheduled. Party A shall hand over pick-up paperwork to Party B for anything that needs Party B to pick up, and Party B shall be responsible for the cost of transportation. In the event of materials and equipment provided by Party A are poor in quality or have difference in technical specifications, and have incurred losses to the project, Party A shall be responsible for the losses. Having accepted the materials provided by Party A, Party B shall be responsible for the storage and protection of the materials. Party B shall be responsible for the compensation of any loss as a result of improper protection by Party B or theft.

9.2 本工程一般材料、设备原则上由乙方负责采购供应、场内运输及现场保管、就位等。对所选用的材料和设备的品牌及产地，乙方应事先征得甲方认可方可采购。电器材料配件、设备应优先采用名牌产品或中外合资产品。甲方有权在合同价格范围内指定部分材料采购，但不能免除乙方对工程质量、安全、成品保护、保修等的相关责任。所有材料必须是全新的。

In principle Party B should be responsible for the acquisition, on-site transfer, protection, and positioning of general materials and equipment of this project. Brands and manufacturer of the materials and equipment should be confirmed by Party A prior to purchase. Famous brands or products from Joint Venture vendors are preferred for electric appliances and accessories. Party A has the right to appoint pat of the materials within the contract price range, but Party B's liabilities of project quality, safety, protection of finished goods, and warranty should NOT be exempted. All materials must be brand new.

9.3 如乙方无法满足9.2条款要求时，甲方可按9.1条款办理。

In the event of Party B fails to meet the requirement of clause 9.2, Party A may process the issue according to clause 9.1.

9.4 凡由乙方采购的材料、设备，如不符合相关质量要求或规格有差异，应立即停止使用。若已使用，对工程造成的损失由乙方负责。

Any materials and equipment purchased by Party B that can't meet relevant quality standards or have difference in specifications must be stopped to use. If such materials or equipment have already been used, Party B shall be responsible for the loss.

9.5　甲方在招标文件内指定材料品牌的，乙方必须按照甲方指定的品牌进行采购并施工，乙方没有权利私自使用其他品牌进行替代。

Party B shall purchase the materials and equipment that Party A has appointed according to the tender documents and do the construction and installation accordingly. Party B has NO right to replace these materials and equipment with other brands without the consent of Party A.

第十条　变更
Article 10 Variations

10.1　在工程进行中，甲方代表随时有权利对图纸和说明提出更改、增加、替换或缩减的要求，但该变更须经甲方代表以书面形式确认方能有效。

During the process of the project, the representative of Party A has the right to request for variation, addition, substitution, and reduction on the drawings and explanations at any time, but such variations shall only be regarded as effective with the confirmation of the representative of Party A in writing.

第十一条　有关安全生产和防火的约定
Article 11 Agreement on Safety and Fire Control in Construction

11.1　甲方提供的施工图纸或作法说明，应符合《中华人民共和国消防条例》中的有关防火设计规范。

Construction drawings and Explanations of Work Procedures provided by Party A should meet all standards of fire control design set forth by the Fire Code of the People's Republic of China.

11.2　乙方在施工期间应严格遵守《建筑安装工程安全技术规程》《建筑安装工人安全操作规程》《中华人民共和国消防条例》和其他相关的法规、规范。

Party B must strictly adhere to the Safety and Technical Standards of Construction and Infrastructure Installation, Safety Standards for Construction Workers, the Fire Code of the People's Republic of China, and other relevant regulations and standards.

11.3　由于乙方在施工生产过程中违反有关安全操作规程、消防条例，导致发生安全或火灾事故，乙方应承担由此引发的一切经济损失。发生事故后，乙方应立即上报政府有关部门并通知甲方代表，同时按政府有关部门要求处理。乙方对事故负全责，并确保甲方完全免责，甲方不承担任何财务或非财务的责任。

In the event of accidents regarding safety or fire control occur due to Party B's violation of safety standards and fire code during construction, Party B shall be fully responsible for all financial losses. Upon accident happened Party B should promptly report the event to relevant government authorities and inform the representative of Party A, and should handle the issue according to government instructions. Party B shall be fully responsible for such accidents, and should guarantee that Party A is fully exempted from liabilities, and Party A shall NOT be responsible for any financial or non-financial liabilities.

11.4 乙方在开工前应提出安全措施,甲方代表有权阻止或辞退违反安全措施的乙方雇员或乙方分包商的雇员。乙方无条件同意替换雇员。

Party B should submit safety measures prior to the beginning of the project; the representative of Party A has the right to stop or dismiss employees of Party B or its subcontractors who have violated the safety measures. Party B should agree to replace such employees unconditionally.

11.5 乙方在需要动用明火作业时,应按规定向甲方审批,经批准后才可施工。

In the event of Party B needs to use open fire for operation, Party B should request consent from Party A, and can ONLY proceed with the approval of Party A.

11.6 乙方需与每个施工人员签订安全协议,并做好上岗前的安全教育工作。

Party B should sign safety agreement with each individual construction worker, and provide proper education on safety prior to first day of work.

11.7 室外施工场地和垃圾淤泥堆放场地必须限制在允许范围内,外围做实用美观的围挡防护。

Outdoor construction space and storage space for wastes should be restricted within the permitted region, and should be enclosed with decent fences for protection.

11.8 施工现场不准做饭和住宿,只允许适当人员看守现场。

No cooking and lodging is allowed on the construction site. Only authorized security personnel are permitted to stay at the site over night.

11.9 施工人员按规定通道使用指定卫生间,不准在卫生间内洗澡等。

Construction personnel should use toilets according to the approved passageway, and are NOT allowed to take showers in the restrooms.

第12条 转让与分包

Article 12 Contract Transfer and Subcontracting

12.1 没有甲方事先的书面同意,乙方不得转让本合同的全部或其部分或其任何利益。

Party B shall NOT transfer the contract and any part of the contract and its corresponding to any third party without Party A's consent in writing.

12.2 除非合用另有规定,乙方未经甲方代表之事先的书面同意,不得分包工程的任何部分,甲方代表的这种同意,并不得视为免除乙方在本合同项下的任何责任和义务。乙方应当为其分包商及其代理人、雇员、工人的所有行为、违约和疏忽而承担责任,这种责任应等同于其自身或其代理人、雇员、工人之行为、违约和疏忽所致责任。一般计件工作将不被视为本条款所述的分包。甲方代表拥有完全权利在工程开工期间向乙方随时提供为保证工程之适当、充分进行及保养所必需的进一步的详图及指导。乙方应执行并受此约束。

Unless otherwise specified, Party B shall NOT subcontract any part of the project without Party A's consent in writing. The consent of subcontract by the representative of Party A shall NOT be regarded as an exempt of Party B's any responsibilities or liabilities under this contract. Party B shall be held responsible for all behaviors, violations, and negligence of its subcontractors or their agents, employees, or workers, and such responsibility should be equivalent to that incurred by the behaviors, violations, and negligence of Party B itself or its agents, employees, and workers. Usually the piecework is not regarded as subcontracting described in this clause. The representative of Party A has full authority to provide Party B with any additional detailed drawings and instructions necessary to guarantee that the project can proceed and be maintained properly and sufficiently Party B should act accordingly and is subject to the requirement of this authority.

第13条 违约责任
Article 13 Liabilities for Breach

13.1 甲方的违约责任
Party A's liabilities for breach

如甲方未能按合同约定的方式支付合同价款，则甲方应向乙方支付滞纳金；滞纳金自工程款到期日的第二日起计算。每日滞纳金为到期款的0.2%。如由于甲方原因造成付款延期超过25日的，则乙方有权解除本合同，由此造成的损失由甲方承担。

In the event of Party A fails to make payment according to the payment terms, Party A should pay a late fee to Party B; the late fee shall be calculated starting from the second day of the due date for the corresponding payment. Daily rate of the late fee is 0.2% of the corresponding payment. If the delay of the payment is caused by Party A and is delayed for more than 25 days, then Party B can terminate this contract and Party A shall take full responsibility for any loss caused by this.

13.2 乙方的违约责任
Party B's liabilities for breach

13.2.1 如乙方未能按照合同约定时间按期完工的，每逾期一天，乙方向甲方支付工程总价款的0.2%作为违约金。如由于乙方原因造成工程延期超过25日的，则甲方有权解除本合同，由此造成的损失由乙方承担。

In the event of Party B fails to complete the project according to the construction period in the contract, Party B shall pay 0.2% of the total contract amount for each day overdue as penalty for the breach. If the delay of the construction completion is caused by Party B and is delayed for more than 25 days, then Party A can terminate this contract and Party B shall take full responsibility for any loss caused by this.

13.2.2　根据甲方要求，乙方应为其所购买的工程材料向甲方提交材料的合格证以及质量检测报告。如果乙方提供的材料与合格证和质量检测报告不一致，例如价高质劣，以次充好，或假冒伪劣，乙方将被处以罚金，罚金将是工程总价款的30%。

Party B should submit to Party A the QA certificate and Report of Quality Assurance Test for the materials it has purchased; in the event of the materials provided by Party B differs from the QA certificate and the Report of Quality Assurance Test, for instance, products with poor quality and high price, inferior materials faked as superior ones, or counterfeited products, Party B shall be fined at the rate of 30% of the total contract amount.

13.2.3　如乙方未能按照本合同要求购买、使用、安装建设设备、材料，由此对甲方造成的损失，将由乙方采取补救措施，同时赔偿给甲方带来的经济损失。

In the event of Party B fails to purchase, use, install, and build with materials and equipment as described by this contract, Party B shall take remedial measures in response to the loss to Party A, and shall compensate Party A for the financial losses incurred.

第十四条　争议处理

Article 14 Dispute Settlement

14.1　当合同文件内容含糊不清或不相一致时，在不影响工程正常进行的情况下，由发包人承包人协商解决。

In the event that the text of the contract is vague or inconsistent, the Parties shall attempt settlement through friendly negotiation on the premise that the normal progress of the construction project is not affected.

14.2　由本合同引发的争议，应由双方友好协商解决，协商不成的，任何乙方均可将争议提交位于北京的中国国际经济贸易仲裁委员会仲裁。仲裁裁决应是终局的并对双方都具有约束力。仲裁费用的承担由仲裁庭决定。

Any dispute arising from or in connection with this Contract shall be settled through friendly negotiation between the Parties, and in the event of the dispute can NOT be settled through negotiation, either party may submit the dispute to the China International Economic and Trade Arbitration Commission located in Beijing for arbitration which shall be conducted in accordance with the Commission's arbitration rules in effect at the time of applying for arbitration. The arbitral award is final and binding upon both parties. The cost of arbitration shall be burdened according to the Commission's decision.

第15条　不可抗力
Article 15 Force Majeure

15.1　由于不可抗力诸如火灾、水灾、政府强令措施以及其他不可抗力的原因，合同不能履行的，则合同的责任义务将予以延缓。当不可抗力消失，合同将继续履行，合同的责任义务将按不可抗力所造成迟延履行的期间予以同期顺延。

If the contract cannot be fulfilled due to force Majeure, such as fire, flood, government forces and other force majeure factors, the fulfillment of obligations may be delayed. When the force majeure has disappeared, the contract shall be continued to be fulfilled, and the contract obligations shall be extended by the same period of delay caused by the force majeure.

15.2　告知不可抗力方应书面知会另一方事件的发生，同时提供适当的证明条件以及持续时间。另外，告知方应尽力减轻可能给对方造成的损失。

In case of force majeure, the affected party shall notify the other party of the occurrence and the evidence of existence of and the duration of the force majeure. In addition, the disclosing party shall endeavor to reduce losses possibly inflicted to the other party.

15.3　一旦不可抗力事件发生，双方应即时商谈并寻求公正的解决方法，努力减少损失。如该不可抗力情形持续超过二十天，则甲方有权决定解除合同。

In case of force majeure, the two Parties shall discuss in a timely manner in order to seek fair and square solutions and reduce possible losses as much as possible. In the event that the said "Force Majeure" cause lasts over 20 days, Party A has the right to terminate the contract.

第16条　使用已完成工程部分
Article 16 Use of Finished Parts of the Project

16.1　甲方有权利使用任何已经完成部分的工程。
Party A has the right to use any finished pads of the project.

第17条　通知
Article 17 Notices

有关本合同的所有通知应视为已于下列时间送达：
All notices with regard to the contract shall be regarded as delivered at the following times:
(a) 如以专人递送，到达指定地址时；
Delivery by specified personnel, upon arrival at the specified location;
(b) 如以传真方式，发件人的传真机打出成功传送的确认条时；
Delivery by facsimile transfer, upon successful printing of the confirmation notice by the

sender's facsimile machine;

(c) 如以快递方式，在发件后的第三日；
Delivery by courier service, the third day after sending out the notice；

上述通知应送至如下列所接收通知的地址或合同任一方于其后指定的地址。
All above notices should be delivered to the recipient's address listed as follows, or to an address that either Party may appoints later.

乙方代表：××××
Representative of Party B：××××
地址：××××
Address：××××

甲方代表：××××
Representative of Party A：××××
地址：××××
Address：××××

第18条　未尽事宜
Article 18 Miscellaneous

18.1　合同未尽事宜由双方协商签订书面补充协议。补充协议与本合同具有同等效力。

Matters not mentioned herein shall be settled by supplemental agreements by the two Parties. The supplemental agreements shall be equally effective as this contract.

第19条　合同生效
Article 19 Effectively Clause

本合同以中英文书写。一式四份，双方各持两份，签字盖章后生效。如中英文相冲突的，以中文为准。

This contract is written in both Chinese and English in quadruplicate, with both texts being equally authentic and each Party shall hold two copies. The contract becomes effective from the date of signing and stamping by both Parties. In the event of any conflict between the English and Chinese, the latter shall prevail.

甲方: 乙方:××××
Party A:

                                        委托代表人:
                                        Representative:

签订日期: 签订日期:
Date: Date:
                                        Party B:××××

Appendix VI：中华人民共和国建筑法

CONSTRUCTION LAW OF THE PEOPLE'S REPUBLIC OF CHINA

第一章　总　则

CHAPTER I GENERAL PROVISIONS

第一条　为了加强对建筑活动的监督管理，维护建筑市场秩序，保证建筑工程的质量和安全，促进建筑业健康发展，制定本法。

Article 1 With a view to strengthening supervision and regulation of construction activities, maintaining construction market order, ensuring the quality and safety of construction projects and promoting the healthy development of construction industry, this law is hereby formulated.

第二条　在中华人民共和国境内从事建筑活动，实施对建筑活动的监督管理，应当遵守本法。本法所称建筑活动，是指各类房屋建筑及其附属设施的建造和与其配套的线路、管道、设备的安装活动。

Article 2 Construction activities and supervision of construction activities conducted within the territory of the People's Republic of China shall abide by this law.

第三条　建筑活动应当确保建筑工程质量和安全，符合国家的建筑工程安全标准。

Article 3 Construction activities have to guarantee the quality and safety of construction projects and comply with the State's safety standards for construction projects.

第四条　国家扶持建筑业的发展，支持建筑科学技术研究，提高房屋建筑设计水平，鼓励节约能源和保护环境，提倡采用先进技术、先进设备、先进工艺、新型建筑材料和现代管理方式。

Article 4 The State assists the development of construction industry, supports research on construction science and technology in order to raise the level of housing and construction design, encourages saving of energy and environmental protection and advocates the adoption of advanced technology, advanced equipment, advanced process, new building materials and modern managerial methods.

第五条　从事建筑活动应当遵守法律、法规，不得损害社会公共利益和他人的合法权益。任何单位和个人都不得妨碍和阻挠依法进行的建筑活动。

Article 5 Conduct of construction activities shall observe laws and regulations and should in no way undermine social and public interests as well as the legitimate interests of others.

第六条　国务院建设行政主管部门对全国的建筑活动实施统一监督管理。

Article 6 The competent construction administrative department under the State Council shall exercise unified supervision and regulation over construction activities of the whole country.

第二章　建筑许可

CHAPTER II CONSTRUCTION LICENSING

第一节　建筑工程施工许可

SECTION I WORKING LICENSES FOR CONSTRUCTION PROJECTS

第七条　建筑工程开工前，建设单位应当按照国家有关规定向工程所在地县级以上人民政府建设行政主管部门申请领取施工许可证；但是，国务院建设行政主管部门确定的限额以下的小型工程除外。

Article 7 Before the start of construction projects, construction units shall, in accordance with the relevant provisions of the State, apply to the competent construction administrative departments under the prefecture-county governments or above for construction licenses, except for small projects below the threshold value set by the competent construction administrative department under the State Council.

按照国务院规定的权限和程序批准开工报告的建筑工程，不再领取施工许可证。

Construction projects which have obtained approval of work start reports in accordance with the power limits and procedure stipulated by the State Council are no longer required to apply for construction licenses.

第八条　申请领取施工许可证，应当具备下列条件：

Article 8 The following conditions are required for the application of construction licenses:

（一）已经办理该建筑工程用地批准手续；

（1）Having gone through the approval formalities for construction project land use;

（二）在城市规划区的建筑工程，已经取得规划许可证；

（2）Construction projects within urban planned districts have obtained licenses of planning;

（三）需要拆迁的，其拆迁进度符合施工要求；

（3）Where demolition and relocation are necessary, the progress of demolition and relocation comply with the requirements of construction;

（四）已经确定建筑施工企业；

（4）Constructing enterprises for the projects have been chosen;

（五）有满足施工需要的施工图纸及技术资料；

（5）Working drawings and technical data are available to meet the need of construction;

（六）有保证工程质量和安全的具体措施；

（6）Specific measures are available for ensuring the quality and security of construction;

（七）建设资金已经落实；

（7）Funds of construction are available;

（八）法律、行政法规规定的其他条件。

（8）Other conditions as stipulated by laws and administrative regulations.

建设行政主管部门应当自收到申请之日起十五日内，对符合条件的申请颁发施工许可证。

The competent construction administrative departments shall issue construction licenses to qualified applicants within 15 days from the date of receipt of applications.

第九条 建设单位应当自领取施工许可证之日起三个月内开工。因故不能按期开工的，应当向发证机关申请延期；延期以两次为限，每次不超过三个月。既不开工又不申请延期或者超过延期时限的，施工许可证自行废止。

Article 9 Units in charge of construction should start to build the projects within three months since obtaining construction licenses. If work cannot start on schedule, applications should be filed to the license issuing agencies for delay, which shall only be permitted twice in maximum for each case, each permission of delay covering a maximum period of three months. In cases of those projects that neither start on schedule nor apply for delay or exceed the period of permitted delay, the validity of the working licenses automatically expires.

第十条 在建的建筑工程因故中止施工的，建设单位应当自中止施工之日起一个月内，向发证机关报告，并按照规定做好建筑工程的维护管理工作。

Article 10 For projects under construction that come to a halt due to some reasons, the constructing units should report to the license issuing agencies within one month from the stoppage of construction and should also well maintain and manage the construction projects in accordance with provisions.

建筑工程恢复施工时，应当向发证机关报告。中止施工满一年的工程恢复施工前，建设单位应当报发证机关核验施工许可证。

When construction work resumes, reports should be filed with the license issuing agencies. Before the resumption of work on construction projects which have stopped for over one year, units in charge of construction shall apply to the license issuing agencies for verification of their working licenses.

第十一条 按照国务院有关规定批准开工报告的建筑工程，因故不能按期开工或者中止施工的，应当及时向批准机关报告情况。因故不能按期开工超过六个月的，应当重新办理开工报告的批准手续。

Article 11 For construction projects with approval of work-start reports in light of the relevant provisions of the State Council which fail to start construction on schedule or suspend the construction work due to certain reasons, the constructors should report the cases timely to the approving authorities. If the period of non-start of work due to certain reasons exceeds six months, the constructors shall re-go through the formalities of approval for work-start reports.

第二节 从业资格
SECTION II QUALIFICATIONS OF CONSTRUCTORS

第十二条 从事建筑活动的建筑施工企业、勘察单位、设计单位和工程监理单位，应当具备下列条件：

Article 12 Construction engineering enterprises, prospecting units, design units and project supervisory units which engage in construction activities shall possess the following conditions:

（一）有符合国家规定的注册资本；

（1）Possession of registered capital that complies with the State provisions;

（二）有与其从事的建筑活动相适应的具有法定执业资格的专业技术人员；

（2）Possession of specialized technical personnel with statutory professional qualifications consistent with the construction activities they engage in;

（三）有从事相关建筑活动所应有的技术装备；

（3）Possession of technologies and equipment that are necessary for related construction activities;

（四）法律、行政法规规定的其他条件。

（4）Other conditions as stipulated by laws and administrative regulations.

第十三条 从事建筑活动的建筑施工企业、勘察单位、设计单位和工程监理单位，按照其拥有的注册资本、专业技术人员、技术装备和已完成的建筑工程业绩等资质条件，划分为不同的资质等级，经资质审查合格，取得相应等级的资质证书后，方可在其资质等级许可的范围内从事建筑活动。

Article 13 Construction engineering enterprises, prospecting units, design units and project supervisory units are rated into different classes according to their registered capital, professionals and technicians, technical equipment and performance record of completed construction projects, etc. Only when they pass qualification examinations and obtain appropriate qualification certificates may they engage in construction activities commensurate to the scope of their qualification classes.

第十四条 从事建筑活动的专业技术人员，应当依法取得相应的执业资格证书，并在执业资格证书许可的范围内从事建筑活动。

Article 14 Specialized technical personnel engaging in construction activities shall attain the relevant certificates of professional qualification and conduct building activities within the scope of their certificates of professional qualifications.

第三章 建筑工程发包与承包
CHAPTER III CONTRACT AWARDING AND CONTRACTING OF CONSTRUCTION PROJECTS

一般规定
SECTION I GENERAL PROVISIONS

第十五条　建筑工程的发包单位与承包单位应当依法订立书面合同，明确双方的权利和义务。

Article 15 Contract awarding units and contractors for construction projects shall conclude written covenants, specifying the rights and obligations of each party.

发包单位和承包单位应当全面履行合同约定的义务。不按照合同约定履行义务的，依法承担违约责任。

Contract awarding units and contractors should fully fulfill their contractual obligations. Failure to fulfill contractual obligations is subject to contract-breach responsibilities under the laws.

第十六条　建筑工程发包与承包的招标投标活动，应当遵循公开、公正、平等竞争的原则，择优选择承包单位。

Article 16 The tendering and bidding activities for contracting of construction projects shall follow the principle of openness, justice and equal competition, and contractors should be selected based on their merits.

建筑工程的招标投标，本法没有规定的，适用有关招标投标法律的规定。

Tendering of and bidding for construction projects which is not covered by this law shall follow the relevant laws on tendering and bidding.

第十七条　发包单位及其工作人员在建筑工程发包中不得收受贿赂、回扣或者索取其他好处。

Article 17 Contract awarding units and their staff shall not take bribery, commissions or seek other benefits in the process of contract awarding for construction projects.

承包单位及其工作人员不得利用向发包单位及其工作人员行贿、提供回扣或者给予其他好处等不正当手段承揽工程。

Contractors and staff members thereof shall not undertake projects through illegitimate means such as offering bribes, commissions or other benefits to the units contracting out the projects or staff members thereof.

第十八条　建筑工程造价应当按照国家有关规定，由发包单位与承包单位在合同中约定。公开招标发包的，其造价的约定，须遵守招标投标法律的规定。

Article 18 Construction cost of the projects shall be agreed upon in the contracts by both the contractors and the out-contracting units in accordance with relevant regulations of the State. For projects subject to public bidding, the construction cost shall be agreed upon in line with the provisions of the tendering law.

发包单位应当按照合同的约定，及时拨付工程款项。

The out-contracting units shall appropriate project funds in a timely manner in accordance with the provisions of the contracts.

第二节　发包

SECTION II CONTRACT AWARDING

第十九条　建筑工程依法实行招标发包，对不适于招标发包的可以直接发包。

Article 19 Construction projects shall be contracted through public bidding according to law and those not suitable for bidding can be contracted out directly.

第二十条　建筑工程实行公开招标的，发包单位应当依照法定程序和方式，发布招标公告，提供载有招标工程的主要技术要求、主要的合同条款、评标的标准和方法以及开标、评标、定标的程序等内容的招标文件。

Article 20 For construction projects subject to open bidding, the out-contracting units shall, in accordance with legal procedures and methods, publish invitations for bid and provide tender documents containing major technical requirements of the projects subject to bidding, key contract terms, standards and methods for evaluation of bids as well as the procedures for the opening, evaluation and determination of bids.

开标应当在招标文件规定的时间、地点公开进行。开标后应当按照招标文件规定的评标标准和程序对标书进行评价、比较，在具备相应资质条件的投标者中，择优选定中标者。

The bids shall be opened at the time and place specified in the tender document, after which the bids shall be evaluated and compared in accordance with the evaluation standards and procedures specified by the tender documents. Successful bidder shall be selected among the bidders with appropriate qualifications based on their merits.

第二十一条　建筑工程招标的开标、评标、定标由建设单位依法组织实施，并接受有关行政主管部门的监督。

Article 21 The opening, evaluation and determination of bids in the process of tendering for construction projects shall be organized and implemented by the project owners and shall be subject to the supervision of relevant administrative departments.

第二十二条　建筑工程实行招标发包的，发包单位应当将建筑工程发包给依法中标的承包单位。建筑工程实行直接发包的，发包单位应当将建筑工程发包给具有相应资质条件的承包单位。

Article 22 For construction projects subject to bidding, the out-contracting units shall contract out the projects to the contracting units which have won the bid in accordance with law. For construction projects subject to direct contracting, the out-contracting units shall contract out the project to the contracting units with appropriate qualifications.

第二十三条　政府及其所属部门不得滥用行政权力，限定发包单位将招标发包的建筑工程发包给指定的承包单位。

Article 23 The government and its subordinate departments shall not abuse their administrative power by requiring the out-contracting units to contract out the construction projects to designated contracting units.

第二十四条　提倡对建筑工程实行总承包，禁止将建筑工程肢解发包。

Article 24 Overall contracting of projects is encouraged and the contracting of construction projects in parts shall be prohibited.

建筑工程的发包单位可以将建筑工程的勘察、设计、施工、设备采购一并发包给一个工程总承包单位，也可以将建筑工程勘察、设计、施工、设备采购的一项或者多项发包给一个工程总承包单位；但是，不得将应当由一个承包单位完成的建筑工程肢解成若干部分发包给几个承包单位。

The out-contracting units of construction projects can contract out either the survey, design, construction and equipment purchase of the construction units to a single overall contracting unit or one or some of these tasks to an overall contracting unit. However, the out-contracting units shall not divide the construction projects which shall be undertaken by one contracting units into several parts for out-contracting to several different contracting units.

第二十五条 按照合同约定，建筑材料、建筑构配件和设备由工程承包单位采购的，发包单位不得指定承包单位购入用于工程的建筑材料、建筑构配件和设备或者指定生产厂、供应商。

Article 25 For building materials, components and equipment which shall be purchased by the project contracting units as specified by the contracts, the out-contracting units shall not designate the building materials, components and equipment for the projects to be purchased by the contracting units or designate manufacturers and suppliers thereof.

第三节 承包

SICTION Ⅲ CONTRACTING

第二十六条 承包建筑工程的单位应当持有依法取得的资质证书，并在其资质等级许可的业务范围内承揽工程。

Article 26 The project contracting units shall hold legally-obtained certificates of qualifications and contract projects within the business scope as allowed by their level of qualification.

禁止建筑施工企业超越本企业资质等级许可的业务范围或者以任何形式用其他建筑施工企业的名义承揽工程。禁止建筑施工企业以任何形式允许其他单位或者个人使用本企业的资质证书、营业执照，以本企业的名义承揽工程。

Project contracting units are prohibited from contracting projects beyond the business scope as allowed by their level of qualification or in the name of other construction enterprises in any form. Construction enterprises are prohibited from allowing other units or individuals to use their own certificates of qualification, business licenses or their name in any form in order to contract construction projects.

第二十七条 大型建筑工程或者结构复杂的建筑工程，可以由两个以上的承包单位联合共同承包。共同承包的各方对承包合同的履行承担连带责任。

Article 27 Large construction projects or complex construction projects can be jointly contracted by two or more contracting units. Parties to joint project contracting shall bear joint responsibilities for the implementation of the contract.

两个以上不同资质等级的单位实行联合共同承包的，应当按照资质等级低的单位的业务许可范围承揽工程。

For joint contracting by two or more units with different levels of qualification, projects shall be contracted within the business scope of the contracting unit with the lower level of qualification.

第二十八条 禁止承包单位将其承包的全部建筑工程转包给他人，禁止承包单位将其承包的全部建筑工程肢解以后以分包的名义分别转包给他人。

Article 28 Contracting units are prohibited from subcontracting the entire construction projects undertaken by the contracting units to others and they are prohibited from dividing the entire construction projects into parts and then subcontracting those parts to others.

第二十九条 建筑工程总承包单位可以将承包工程中的部分工程发包给具有相应资质条件的分包单位；但是，除总承包合同中约定的分包外，必须经建设单位认可。施工总承包的，建筑工程主体结构的施工必须由总承包单位自行完成。

Article 29 The overall contracting units can subcontract part of the projects to subcontractors with appropriate qualifications. However, except for the subcontracting as specified in the overall contracts, approval from project owners shall be obtained. With regard to overall project contracting, the construction of the main structure of the project shall be completed by the overall contracting units.

建筑工程总承包单位按照总承包合同的约定对建设单位负责；分包单位按照分包合同的约定对总承包单位负责。总承包单位和分包单位就分包工程对建设单位承担连带责任。

The overall contracting units shall be held responsible to project owners in accordance with the provisions of the overall contracts; The subcontractors shall be held responsible to the overall contractors in accordance with the terms of the subcontracting contracts. The overall contractors and subcontractors shall bear joint responsibilities to project owners for the subcontracted projects.

禁止总承包单位将工程分包给不具备相应资质条件的单位。禁止分包单位将其承包的工程再分包。

The overall contracting units are prohibited from subcontracting the projects to units without appropriate qualifications. The subcontractors are prohibitied from re-subcontracting the sub-contracted projects.

第四章 建筑工程监理

CHAPTER IV SUPERVISION OF CONSTRUCTION PROJECTS

第三十条 国家推行建筑工程监理制度。

Article 30 The State adopts the system of project construction supervision.

国务院可以规定实行强制监理的建筑工程的范围。

The State Council can stipulate the scope of construction projects subject to compulsory supervision.

第三十一条　实行监理的建筑工程，由建设单位委托具有相应资质条件的工程监理单位监理。建设单位与其委托的工程监理单位应当订立书面委托监理合同。

Article 31 Construction projects subject to supervision shall be supervised by project supervising units with appropriate qualifications entrusted by the project owners. The project owners shall enter into written contracts on supervision with the entrusted project supervising units.

第三十二条　建筑工程监理应当依照法律、行政法规及有关的技术标准、设计文件和建筑工程承包合同，对承包单位在施工质量、建设工期和建设资金使用等方面，代表建设单位实施监督。

Article 32 The construction supervising units shall, on behalf of project owners, carry out supervision of the construction quality, construction period and the use of construction funds in accordance with laws, administrative regulations and relevant technical standards, design documents and contracts for project contracting.

工程监理人员认为工程施工不符合工程设计要求、施工技术标准和合同约定的，有权要求建筑施工企业改正。

When project supervising personnel think the construction work is not in line with the requirements of project design, technical standards for construction and the terms of the contracts, they have the right to demand corrections from the construction units.

工程监理人员发现工程设计不符合建筑工程质量标准或者合同约定的质量要求的，应当报告建设单位要求设计单位改正。

When project supervising personnel find that project design is not in conformity with the quality standards for project construction or the quality requirements as specified in the contracts, they shall report to the project owners who shall then demand corrections from the designing units.

第三十三条　实施建筑工程监理前，建设单位应当将委托的工程监理单位、监理的内容及监理权限，书面通知被监理的建筑施工企业。

Article 33 Prior to the implementation of the supervision of project construction, the project owners shall inform the construction enterprises in written form of the project supervising unites entrusted, contents for supervision and the scope for supervision.

第三十四条　工程监理单位应当在其资质等级许可的监理范围内，承担工程监理业务。

Article 34 The construction supervising units shall undertake project supervising business within the scope of supervision as allowed by their level of qualifications.

工程监理单位应当根据建设单位的委托，客观、公正地执行监理任务。

The construction supervising units shall carry out construction supervision in an objective and fair manner in accordance with the entrustment by the project owners.

工程监理单位与被监理工程的承包单位以及建筑材料、建筑构配件和设备供应单位不得有隶属关系或者其他利害关系。

The construction supervising units shall not have subordinate relations with or other stakes in the contracting units under supervision and the suppliers of building materials, components and equipment.

工程监理单位不得转让工程监理业务。

The construction supervising units shall not transfer the supervising business.

第三十五条 工程监理单位不按照委托监理合同的约定履行监理义务，对应当监督检查的项目不检查或者不按照规定检查，给建设单位造成损失的，应当承担相应的赔偿责任。

Article 35 If the construction supervising units do not fulfill their obligations in accordance with the terms of the contract for supervision and do not inspect the items that should be supervised or do not follow the regulations in carrying out the supervision, which has resulted in losses on the part of project owners, the supervising units shall bear the corresponding liability for damages.

工程监理单位与承包单位串通，为承包单位谋取非法利益，给建设单位造成损失的，应当与承包单位承担连带赔偿责任。

If the supervising units collude with the contracting units in order to reap illegitimate profits for the contracting units, which has resulted in losses on the part of project owners, the supervising units shall bear joint liability for damages.

第五章 建筑安全生产管理

CHAPTER V MANAGEMENT OF CONSTRUCTION SAFETY AND OPERATION

第三十六条 建筑工程安全生产管理必须坚持安全第一、预防为主的方针，建立健全安全生产的责任制度和群防群治制度。

Article 36 The administration of safety in construction operation shall follow the principle of safety and prevention first. The responsibility system for operation safety and the regime of prevention and treatment by the masses shall be established and perfected.

第三十七条 建筑工程设计应当符合按照国家规定制定的建筑安全规程和技术规范，保证工程的安全性能。

Article 37 The design of construction projects shall be in line with the safety standards for construction and technical specifications in order to ensure the safety performance of the projects.

第三十八条 建筑施工企业在编制施工组织设计时，应当根据建筑工程的特点制定相应的安全技术措施；对专业性较强的工程项目，应当编制专项安全施工组织设计，并采取安全技术措施。

Article 38 When planning construction organization and designing, the construction projects shall formulate appropriate safety technical measures according to the characteristics of the construction projects; With regard to construction projects with a high degree of technical speciality, special safety construction organization and designs shall be formulated and safety technical measures shall be adopted.

第三十九条 建筑施工企业应当在施工现场采取维护安全、防范危险、预防火灾等措施；有条件的，应当对施工现场实行封闭管理。

Article 39 The construction enterprises shall take measures aimed at maintaining safety, preventing dangers and fires. When condition allows, the construction site shall be sealed up.

施工现场对毗邻的建筑物、构筑物和特殊作业环境可能造成损害的，建筑施工企业应当采取安全防护措施。

When the construction site may cause damages to the adjacent buildings, structures and special operating environment, the construction enterprises shall adopt safety and preventive measures.

第四十条 建设单位应当向建筑施工企业提供与施工现场相关的地下管线资料，建筑施工企业应当采取措施加以保护。

Article 40 The project owners shall provide the construction enterprises with the documents concerning underground pipeline layout related to the construction site and the construction enterprises shall take preventive measures.

第四十一条 建筑施工企业应当遵守有关环境保护和安全生产的法律、法规的规定，采取控制和处理施工现场的各种粉尘、废气、废水、固体废物以及噪声、振动对环境的污染和危害的措施。

Article 41 In accordance with laws and regulations concerning environmental protection and safety in operation, construction enterprises shall adopt measures to control and clear environmental pollution and harm resulting from various kinds of dust, waste gas, waste water, solid waste materials, noise and vibration at construction sites.

第四十二条 有下列情形之一的，建设单位应当按照国家有关规定办理申请批准手续：

In accordance with relevant state regulations, units undertaking projects shall go through application and approval procedures for following matters:

（一）需要临时占用规划批准范围以外场地的；

(1) Need to temporarily occupy areas not covered by approved construction programs;

（二）可能损坏道路、管线、电力、邮电通信等公共设施的；

(2) Likely to damage roads, pipes and electric wires, power supply and post and telecommunication equipment and other public facilities;

（三）需要临时停水、停电、中断道路交通的；

(3) Need to temporarily cut off water and power supply or hold up traffic;

（四）需要进行爆破作业的；

(4) Need to conduct blasting operations;

（五）法律、法规规定需要办理报批手续的其他情形。

(5) Other matters for which application and approval procedures are required in

accordance with the stipulations of laws and regulations.

第四十三条 建设行政主管部门负责建筑安全生产的管理，并依法接受劳动行政主管部门对建筑安全生产的指导和监督。

Article 43 Administrative authorities on construction are responsible for the administration of safety in construction operation and shall, according to law, subject themselves to the instruction and supervision on safety in construction operation by administrative authorities on labor.

第四十四条 建筑施工企业必须依法加强对建筑安全生产的管理，执行安全生产责任制度，采取有效措施，防止伤亡和其他安全生产事故的发生。

Article 44 Construction enterprises shall, according to law, strengthen the safety in construction operation, enforce the safe operation responsibility system, and prevent casualties and other operation accidents by adopting effective measures.

建筑施工企业的法定代表人对本企业的安全生产负责。

The legal representative of a construction enterprise is responsible for the safety in operation of this enterprise.

第四十五条 施工现场安全由建筑施工企业负责。实行施工总承包的，由总承包单位负责。分包单位向总承包单位负责，服从总承包单位对施工现场的安全生产管理。

Article 45 Construction enterprises are responsible for the safety at construction sites. As to a construction project under an overall contract, the overall contractor is responsible for the safety at construction sites. Subcontractors shall hold themselves responsible to the overall contractor and be subject to the administration of the overall contractor concerning safe operation at construction sites.

第四十六条 建筑施工企业应当建立健全劳动安全生产教育培训制度，加强对职工安全生产教育的培训；未经安全生产教育培训的人员，不得上岗作业。

Article 46 Construction enterprises shall establish a sound system of education and training in safe operation, strengthen the education and training of their staff members in safe operation. Those without receiving a training in safe operation are prohibited from going on duty.

第四十七条 建筑施工企业和作业人员在施工过程中，应当遵守有关安全生产的法律、法规和建筑行业安全规章、规程，不得违章指挥或者违章作业。作业人员有权对影响人身健康的作业程序和作业条件提出改进意见，有权获得安全生产所需的防护用品。作业人员对危及生命安全和人身健康的行为有权提出批评、检举和控告。

Article 47 In the course of construction operation, construction enterprises and working personnel shall observe laws and regulations concerning safe operation as well as safety rules of the construction industry. They shall not command or operate in violation of relevant regulations. Working personnel are entitled to forward proposals for improving the operating programs and conditions which adversely affect health and to obtain protective equipment as required for safe operation. They are entitled to criticize, report and accuse of actions endangering vital safety and personal health.

第四十八条　建筑施工企业应当依法为职工参加工伤保险缴纳工伤保险费。鼓励企业为从事危险作业的职工办理意外伤害保险，支付保险费。

Article 48 Construction enterprises should pay industrial injury insurance premium for workers in accordance with the law. Encourage enterprises to give accident and casualty insurance to workers engaged in dangerous operations and pay insurance premium for them.

第四十九条　涉及建筑主体和承重结构变动的装修工程，建设单位应当在施工前委托原设计单位或者具有相应资质条件的设计单位提出设计方案；没有设计方案的，不得施工。

Article 49 As to a decoration project involving the change of the main part and bearing structure of the building, the unit undertaking the project shall entrust the original designing institution or other designing institutions of equivalent capability with the responsibility for drawing up the design scheme before the start of construction operation. The project shall not be started without a design scheme.

第五十条　房屋拆除应当由具备保证安全条件的建筑施工单位承担，由建筑施工单位负责人对安全负责。

Article 50 Demolition shall be taken by the construction enterprises being able to ensure safety. The heads of those enterprises are responsible for safety.

第五十一条　施工中发生事故时，建筑施工企业应当采取紧急措施减少人员伤亡和事故损失，并按照国家有关规定及时向有关部门报告。

Article 51 If an accident occurred in the course of construction operation, construction enterprises shall take emergency measures to reduce casualties and losses resulting from the accident and immediately report to the departments concerned in accordance with relevant state regulations.

第六章　建筑工程质量管理

CHAPTER VI MANAGEMENT OF CONSTRUCTION PROJECT QUALITY

第五十二条　建筑工程勘察、设计、施工的质量必须符合国家有关建筑工程安全标准的要求，具体管理办法由国务院规定。

Article 52 The survey, design and construction of a project must meet the requirement of relevant state safety standards on construction projects. The specific administrative measures shall be stipulated by the State Council.

有关建筑工程安全的国家标准不能适应确保建筑安全的要求时，应当及时修订。

The state safety standards on construction projects which can not ensure the safety of buildings shall be amended without delay.

第五十三条　国家对从事建筑活动的单位推行质量体系认证制度。从事建筑活动的单位根据自愿原则可以向国务院产品质量监督管理部门或者国务院产品质量监督管理部门授权的部门认可的认证机构申请质量体系认证。经认证合格的，由认证机构颁发质量体系认证证书。

Article 53 The state shall establish the quality certification system for units engaged in construction. On a voluntary basis, units engaged in construction may, when applying for quality certification, submit their applications to the certification institutions approved by the product quality control departments under the State Council and by other departments authorized by the product quality control departments under the State Council. If certified, the units engaged in construction shall be granted with Quality Certificates by the certification institutions.

第五十四条 建设单位不得以任何理由，要求建筑设计单位或者建筑施工企业在工程设计或者施工作业中，违反法律、行政法规和建筑工程质量、安全标准，降低工程质量。

Article 54 Units which are undertaking projects shall not, with any excuse, demand designing institutions or construction enterprises to act in violation of laws, administrative regulations and quality and safety standards on construction projects and lower the quality of projects in the course of design and construction operation.

建筑设计单位和建筑施工企业对建设单位违反前款规定提出的降低工程质量的要求，应当予以拒绝。

Designing institutions or construction enterprises shall reject the requests from owner for lowering the quality of projects in violation of the stipulations of previous articles.

第五十五条 建筑工程实行总承包的，工程质量由工程总承包单位负责，总承包单位将建筑工程分包给其他单位的，应当对分包工程的质量与分包单位承担连带责任。分包单位应当接受总承包单位的质量管理。

Article 55 As to a construction project under an overall contract, the overall contractor is responsible for the quality of the project. If the overall contractor jobbed out to subcontractors, the overall contractor should assume responsibility with subcontractors for the quality of projects under subcontracts. Subcontractors shall subject themselves to the quality control of the overall contractor.

第五十六条 建筑工程的勘察、设计单位必须对其勘察、设计的质量负责。勘察、设计文件应当符合有关法律、行政法规的规定和建筑工程质量、安全标准、建筑工程勘察、设计技术规范以及合同的约定。设计文件选用的建筑材料、建筑构配件和设备，应当注明其规格、型号、性能等技术指标，其质量要求必须符合国家规定的标准。

Article 56 Survey and designing institutions shall be responsible for the survey and design quality of construction projects. Survey and designing documents shall conform to the stipulations of relevant laws and administrative regulations, quality and safety standards on construction projects, survey and design technical codes on construction projects and contracted terms. The specification, model, function and other specifications of construction materials, structural parts, construction fittings and equipment chosen in design documents shall be specified in detail. Their quality must meet the standards set by the state.

第五十七条 建筑设计单位对设计文件选用的建筑材料、建筑构配件和设备，不得指定生产厂、供应商。

Article 57 Designing institutions shall not designate manufacturers and suppliers of construction materials, structural parts, construction fittings and equipment chosen in design documents.

第五十八条 建筑施工企业对工程的施工质量负责。

Article 58 Construction enterprises are responsible for the quality of construction projects.

建筑施工企业必须按照工程设计图纸和施工技术标准施工,不得偷工减料。工程设计的修改由原设计单位负责,建筑施工企业不得擅自修改工程设计。

Construction enterprises must carry out construction operation in accordance with design drawings of projects and construction technical standards, and shall not cheat on work and materials. The original designing institution shall bear the responsibility for altering the design of projects. Construction enterprises shall not alter the design of projects without authorization.

第五十九条 建筑施工企业必须按照工程设计要求、施工技术标准和合同的约定,对建筑材料、建筑构配件和设备进行检验,不合格的不得使用。

Article 59 Construction enterprises shall examine the quality of construction materials, structural parts, construction fittings and equipment in accordance with the requirement of the design of projects, construction technical standards and contracted terms. Those proved not up to the standards shall not be put into use.

第六十条 建筑物在合理使用寿命内,必须确保地基基础工程和主体结构的质量。

Article 60 Within normal service life time of buildings, the quality of foundations and main parts of buildings shall be guaranteed.

建筑工程竣工时,屋顶、墙面不得留有渗漏、开裂等质量缺陷;对已发现的质量缺陷,建筑施工企业应当修复。

Upon the completion of construction projects, there shall not be leakage, cracking and other defects left on roofs and walls. Construction enterprises shall repair the defects having been found out.

第六十一条 交付竣工验收的建筑工程,必须符合规定的建筑工程质量标准,有完整的工程技术经济资料和经签署的工程保修书,并具备国家规定的其他竣工条件。

Article 61 Construction projects having been completed and accepted through examination shall meet the stipulated quality standards on construction projects, have complete technical and economic data of projects and warranties issued by builders, and satisfy other requirements set out by the state for the completion of construction projects.

建筑工程竣工经验收合格后,方可交付使用;未经验收或者验收不合格的,不得交付使用。

Only after a completed construction project is proved to meet the standards through examination can it be delivered for use. Construction projects having not been examined and accepted or having failed in examination shall not be delivered for use.

第六十二条 建筑工程实行质量保修制度。

Article 62 The system of the warranty of quality shall be established for construction projects.

建筑工程的保修范围应当包括地基基础工程、主体结构工程、屋面防水工程和其他土建工程，以及电气管线、上下水管线的安装工程，供热、供冷系统工程等项目；保修的期限应当按照保证建筑物合理寿命年限内正常使用、维护使用者合法权益的原则确定。具体的保修范围和最低保修期限由国务院规定。

The scope of warranties on construction projects covers foundations, main parts, the leak prevention of roof covering and other building projects as well as the installation of electric wires and gas, water supply and drainage pipes, and heating and air-conditioning systems. The time frames of warranties shall be determined in accordance with the principle of ensuring the regular use of buildings within normal service life time and safeguarding users' legitimate rights and interests. The specific scope and minimum time frame of warranties shall be stipulated by the State Council.

第六十三条 任何单位和个人对建筑工程的质量事故、质量缺陷都有权向建设行政主管部门或者其他有关部门进行检举、控告、投诉。

Article 63 Any unit and individual are entitled to report, accuse of and complain about the quality accidents and defects of construction projects to administrative authorities on construction and other departments concerned.

第七章 法律责任

CHAPTER VII LEGAL RESPONSIBILITIES

第六十四条 违反本法规定，未取得施工许可证或者开工报告未经批准擅自施工的，责令改正，对不符合开工条件的责令停止施工，可以处以罚款。

Article 64 Construction enterprises, which act in violation of the stipulations of this Law to start construction operation without construction permit or at the time when the application for construction operation has not yet been approved, shall be ordered to correct themselves. Construction enterprises of which construction projects can not meet the requirement for starting operation shall be ordered to stop construction operation and may be imposed fine penalties.

第六十五条 发包单位将工程发包给不具有相应资质条件的承包单位的，或者违反本法规定将建筑工程肢解发包的，责令改正，处以罚款。

Article 65 A contract awarder who contracts projects to unqualified contractors or contract projects by dividing the projects into different parts in violation of the stipulations of this Law shall be ordered to correct himself and shall be imposed fine penalties.

超越本单位资质等级承揽工程的，责令停止违法行为，处以罚款，可以责令停业整顿，降低资质等级；情节严重的，吊销资质证书；有违法所得的，予以没收。

An organization which exceeds its level of qualification to contract projects shall be ordered to stop the illegal activities and imposed fine penalties, and can be ordered to stop business operations for rectification, with its qualification level reduced. An organization found to have serious violations shall be revoked business license, with all its illegal incomes confiscated.

未取得资质证书承揽工程的，予以取缔，并处罚款；有违法所得的，予以没收。

An organization which contracts projects without a certificate of qualification shall be outlawed and imposed fine penalties, with all its illegal incomes confiscated.

以欺骗手段取得资质证书的，吊销资质证书，处以罚款；构成犯罪的，依法追究刑事责任。

An organization which has obtained a certificate of qualification through cheating shall be revoked the certificate and imposed fine penalties, and shall be prosecuted for criminal liabilities according to law for any crimes committed.

第六十六条 建筑施工企业转让、出借资质证书或者以其他方式允许他人以本企业的名义承揽工程的，责令改正，没收违法所得，并处罚款，可以责令停业整顿，降低资质等级；情节严重的，吊销资质证书。对因该项承揽工程不符合规定的质量标准造成的损失，建筑施工企业与使用本企业名义的单位或者个人承担连带赔偿责任。

Article 66 A construction enterprise which transfers or lends its certificate of qualification to others or which allows others to contract projects in its name shall be ordered to correct itself and imposed fine penalties, with all illegal incomes confiscated. It can also be ordered to stop business operations and to have its level of qualification reduced. An enterprise found to have serious violations shall be revoked the certificate of qualification. The construction enterprise and the organization or individual which has used the name of the construction enterprise shall assume the associated liabilities for losses incurred from the sub-par quality of the contracted project.

第六十七条 承包单位将承包的工程转包的，或者违反本法规定进行分包的，责令改正，没收违法所得，并处罚款，可以责令停业整顿，降低资质等级；情节严重的，吊销资质证书。

Article 67 A project contractor which subcontracts the contracted project in violation of the stipulations of this Law shall be ordered to correct the situation, imposed fine penalties, have all illegal incomes confiscated, and can be ordered to stop business operations and lower the level of qualification. A contractor found to have serious violations shall be revoked the certificate of qualification.

承包单位有前款规定的违法行为的，对因转包工程或者违法分包的工程不符合规定的质量标准造成的损失，与接受转包或者分包的单位承担连带赔偿责任。

The contractor which has violated the above clause and the subcontractor shall assume the associate liabilities for the losses incurred from the sub-par quality of the project due to subcontracting.

第六十八条 在工程发包与承包中索贿、受贿、行贿，构成犯罪的，依法追究刑事责任；不构成犯罪的，分别处以罚款，没收贿赂的财物，对直接负责的主管人员和其他直接责任人员给予处分。

Article 68 Acts of asking for bribes, taking bribes and offering bribes in the course of contract awarding and project contracting which have committed crimes shall be prosecuted for criminal liabilities. Those who have not committed crimes shall be imposed fine penalties respectively and have the bribes confiscated. The people in direct charge and the people directly involved in the acts shall be imposed punishment.

对在工程承包中行贿的承包单位，除依照前款规定处罚外，可以责令停业整顿、降低资质等级或者吊销资质证书。

A contractor who offers bribes in project contracting shall be ordered to stop business operations for rectification, have the level of qualification reduced or shall be revoked the certificate of qualification in addition to the punishment provided for in the previous clause.

第六十九条 工程监理单位与建设单位或者建筑施工企业串通，弄虚作假、降低工程质量的，责令改正，处以罚款，降低资质等级或者吊销资质证书；有违法所得的，予以没收；造成损失的，承担连带赔偿责任；构成犯罪的，依法追究刑事责任。

Article 69 A project supervisory organization which collides with the project undertaker or the construction enterprise to practice fraud and lower the quality of the project shall be ordered to correct the situation, imposed fine penalties, shall have the level of qualification reduced or be revoked the certificate of qualification, with all illegal incomes confiscated. Where losses are incurred, the organization shall assume associated liabilities for compensation. And the organization found to have committed crimes shall be prosecuted for criminal liabilities according to law.

工程监理单位转让监理业务的，责令改正，没收违法所得，可以责令停业整顿，降低资质等级；情节严重的，吊销资质证书。

A project supervisory organization which transfers the business operation of project supervision shall be ordered to correct the situation, shall have all illegal incomes confiscated, and can be ordered to stop business operations for rectification and have its level of qualification reduced. An organization found to have serious violations shall be revoked the certificate of qualification.

第七十条 违反本法规定，涉及建筑主体或者承重结构变动的装修工程擅自施工的，责令改正，处以罚款；造成损失的，承担赔偿责任；构成犯罪的，依法追究刑事责任。

Article 70 An organization which violates the stipulations of this Law to undertake decoration project by changing the major structure or load-bearing structure of the building shall be ordered to correct the situation and imposed fine penalties; shall assume liabilities of compensation for losses incurred thereof; and shall be prosecuted for criminal liabilities according to law for any crimes committed.

第七十一条　建筑施工企业违反本法规定，对建筑安全事故隐患不采取措施予以消除的，责令改正，可以处以罚款；情节严重的，责令停业整顿，降低资质等级或者吊销资质证书；构成犯罪的，依法追究刑事责任。

Article 71 A construction enterprise which does not adopt necessary measures to eliminate potential dangers of accidents in violation of the stipulations of this Law shall be ordered to correct itself and can be imposed fine penalties. A construction enterprise with serious violations shall be ordered to stop business operations for rectification, have its level of qualification reduced or shall be revoked the certificate of qualification, and shall be prosecuted for criminal liabilities according to law for any crimes committed.

建筑施工企业的管理人员违章指挥、强令职工冒险作业，因而发生重大伤亡事故或者造成其他严重后果的，依法追究刑事责任。

The management of a construction enterprise who gives directions in violation of regulations and forces workers to take risks in performing operations shall be prosecuted for criminal liabilities for major casualties or other serious consequences.

第七十二条　建设单位违反本法规定，要求建筑设计单位或者建筑施工企业违反建筑工程质量、安全标准，降低工程质量的，责令改正，可以处以罚款；构成犯罪的，依法追究刑事责任。

Article 72 A project undertaker which violates the stipulations of this Law to request the project designer and the construction enterprise to alleviate from quality and safety standards so as to lower construction quality shall be ordered to correct the situation, can be imposed fine penalties, and shall be prosecuted for criminal liabilities according to law for any crimes committed.

第七十三条　建筑设计单位不按照建筑工程质量、安全标准进行设计的，责令改正，处以罚款；造成工程质量事故的，责令停业整顿，降低资质等级或者吊销资质证书，没收违法所得，并处罚款；造成损失的，承担赔偿责任；构成犯罪的，依法追究刑事责任。

Article 73 A project designer which does not abide by quality and safety standards in designing projects shall be ordered to correct itself and imposed fine penalties. Where major quality accidents are incurred, the designer shall be ordered to stop business operations for rectification, shall have its level of qualification reduced or shall be revoked the certificate of qualification and imposed fine penalties, with all illegal incomes confiscated. The designer shall assume liabilities of compensation for losses incurred thereof, and shall be prosecuted for criminal liabilities according to law for any crimes committed.

第七十四条　建筑施工企业在施工中偷工减料的，使用不合格的建筑材料、建筑构配件和设备的，或者有其他不按照工程设计图纸或者施工技术标准施工的行为的，责令改正，处以罚款；情节严重的，责令停业整顿，降低资质等级或者吊销资质证书；造成建筑工程质量不符合规定的质量标准的，负责返工、修理，并赔偿因此造成的损失；构成犯罪的，依法追究刑事责任。

Article 74 A construction enterprise which has been found to have scamped work and used inferior materials, unqualified construction materials, parts and components and equipment or have other acts inconsistent with the blueprints or technical standards shall be ordered to correct itself and imposed fine penalties. An enterprise found to have serious violations shall be ordered to stop business operations for rectification, shall have its level of qualification reduced or shall be revoked the certificate of qualification. Where construction quality is found lower than the stipulated quality standards, the said enterprise shall be responsible for the reconstruction and repair, shall compensate the losses incurred thereof and shall be prosecuted for criminal liabilities according to law for any crimes committed.

第七十五条　建筑施工企业违反本法规定，不履行保修义务或者拖延履行保修义务的，责令改正，可以处以罚款，并对在保修期内因屋顶、墙面渗漏、开裂等质量缺陷造成的损失，承担赔偿责任。

Article 75 A construction enterprise which does not perform its obligations pertaining to maintenance or default on these obligations in violation of the stipulations of this Law shall be ordered to correct itself, can be imposed fine penalties and shall assume the liabilities of compensation for losses incurred from quality defects such as ceiling or wall leaking and cracks during the term of maintenance.

第七十六条　本法规定的责令停业整顿、降低资质等级和吊销资质证书的行政处罚，由颁发资质证书的机关决定；其他行政处罚，由建设行政主管部门或者有关部门依照法律和国务院规定的职权范围决定。

Article 76 The administrative punishment provided for in the stipulations of this Law like forced stoppage of business operations for rectification, lowering of the level of qualification and revocation of the certificate of qualification shall be decided by the issuing authority of the certificate of qualification. Other administrative punishment shall be decided by competent authorities of construction or relevant authorities based on the purview provided for in the law or authorized by the State Council.

依照本法规定被吊销资质证书的，由工商行政管理部门吊销其营业执照。

Administrative authorities of industry and commerce shall revoke business licenses of those enterprises whose certificates of qualification have been revoked in accordance with the stipulations of this Law.

第七十七条　违反本法规定，对不具备相应资质等级条件的单位颁发该等级资质证书的，由其上级机关责令收回所发的资质证书，对直接负责的主管人员和其他直接责任人员给予行政处分；构成犯罪的，依法追究刑事责任。

Article 77 Certificates of qualification issued to unqualified organizations in violation of the stipulations of this Law shall be ordered by the superior administrative authority for revocation of the certificates issued. People in direct charge and people directly involved shall be imposed administrative punishment and criminal liabilities shall be prosecuted according to law for any crimes committed.

第七十八条 政府及其所属部门的工作人员违反本法规定，限定发包单位将招标发包的工程发包给指定的承包单位的，由上级机关责令改正；构成犯罪的，依法追究刑事责任。

Article 78 Employees of government agencies or their subordinate authorities who violate the stipulations of this Law to direct contract awarders to contract the bidding project to designated contractors shall be ordered by superior authorities to correct themselves and shall be prosecuted for criminal liabilities according to law for any crimes committed.

第七十九条 负责颁发建筑工程施工许可证的部门及其工作人员对不符合施工条件的建筑工程颁发施工许可证的，负责工程质量监督检查或者竣工验收的部门及其工作人员对不合格的建筑工程出具质量合格文件或者按合格工程验收的，由上级机关责令改正，对责任人员给予行政处分；构成犯罪的，依法追究刑事责任；造成损失的，由该部门承担相应的赔偿责任。

Article 79 Authorities responsible for the issuance of construction permits and their employees who issue construction permits to unqualified construction projects and authorities responsible for the supervision and inspection of project quality or acceptance test and their employees who issue documents of qualification to unqualified projects or accept unqualified projects as qualified ones shall be ordered by superior authorities to correct themselves. People directly involved shall be imposed administrative punishment and shall be prosecuted for criminal liabilities according to law for any crimes committed. Liabilities of compensation shall be assumed by the authorities concerned for any losses incurred thereof.

第八十条 在建筑物的合理使用寿命内，因建筑工程质量不合格受到损害的，有权向责任者要求赔偿。

Article 80 People suffering from inferior construction quality during the reasonable life span of the buildings are entitled to compensation from people responsible for the damages.

第八十一条 本法关于施工许可、建筑施工企业资质审查和建筑工程发包、承包、禁止转包，以及建筑工程监理、建筑工程安全和质量管理的规定，适用于其他专业建筑工程的建筑活动，具体办法由国务院规定。

Article 81 Stipulations of this Law pertinent to construction permit, examination of the qualification of construction enterprises, awarding, contracting and ban on subcontracting of construction projects as well as supervision, safety and quality management of the construction projects shall be applicable to the construction of other special projects, with details formulated by the State Council.

第八十二条 建设行政主管部门和其他有关部门在对建筑活动实施监督管理中，除按照国务院有关规定收取费用外，不得收取其他费用。

Article 82 Apart from charges provided for in relevant regulations of the State Council, competent authorities responsible for construction and other relevant authorities shall not collect other fees while exercising supervision over construction activities.

第八十三条　省、自治区、直辖市人民政府确定的小型房屋建筑工程的建筑活动，参照本法执行。

Article 83 Construction activities of small-sized buildings designated by the people's governments of provinces, autonomous regions and municipalities shall be carried out based on the stipulations of this Law mutatis mutandis.

依法核定作为文物保护的纪念建筑物和古建筑等的修缮，依照文物保护的有关法律规定执行。

Renovation of memorials and ancient buildings certified by law as cultural relics for protection shall be carried out in accordance with relevant laws of protection of cultural relics.

抢险救灾及其他临时性房屋建筑和农民自建低层住宅的建筑活动，不适用本法。

The stipulations of this Law are not applicable to buildings erected for risk and disaster relief, other buildings for provisional use and low-stored residential buildings constructed by farmers for their own use.

第八十四条　军用房屋建筑工程建筑活动的具体管理办法，由国务院、中央军事委员会依据本法制定。

Article 84 Specific regulations for the administration of construction of buildings for military use shall be formulated by the State Council and the Military Committee of the Central Party Committee in accordance with the stipulations of this Law.

参 考 文 献

[1] 郭向荣,陈政清.土木工程专业英语[M].北京:中国铁道工业出版社,2001.

[2] 梁思成.为什么研究中国建筑[M].北京:外语教学与研究出版社,2011.

[3] 张成国.建筑英语口语[M].南京:东南大学出版社,2013.

[4] 何飞云,袁荣儿.建筑行业英语[M].北京:中国水利水电出版社,2011.

[5] 韩薇,张华明.建筑工程专业英语[M].北京:北京大学出版社,2012.

[6] 赵琼梅,胡莹.建筑工程英语[M].北京:中国建筑工业出版社,2012.

[7] 蒋海燕.建筑专业英语[M].北京:中国建材工业出版社,2003.

[8] 孔娟.土建英语基础教程[M].郑州:黄河水利出版社,2013.

[9] 马彩玲.建筑工程英语[M].北京:科学出版社,2012.

[10] 练长城.建筑职业英语[M].天津:天津大学出版社,2011.

[11] 谷素华.行业英语入门——建筑工程及艺术设计类[M].石家庄:河北人民出版社,2011.

[12] 王宝华,张培芳.建筑实用英语[M].大连:大连理工大学出版社,2012.

[13] 颜碧宇,尤晓洁,王细娇.建筑行业英语[M].北京:电子工业出版社,2014.

[14] 蒋春霞.建筑工程专业英语[M].北京:化学工业出版社,2012.

[15] 韦成秀.建筑英语[M].北京:中国建筑工业出版社,1997.

[16] 盛根友.建筑英语[M].2版.北京:中国劳动社会保障出版社,2012.

[17] 盛根友.建筑施工现场英语情景会话[M].北京:中国建筑工业出版社,2010.

[18] 杨芳.土建英语(新职业英语行业篇)[M].北京:外语教学与研究出版社,2010.

[19] http://en.wikipedia.org/wiki/Ancient_Egyptian_architecture

[20] http://en.wikipedia.org/wiki/Green_building

[21] http://wenku.baidu.com/view/c41dff176c175f0e7cd13793.html

[22] http://www.travelchinaguide.com/intro/architecture/styles/

[23] http://wenku.baidu.com